THE COMMISSION OF THE EUROPEAN COMMUNITIES

Denitration of Radioactive Liquid Waste

Proceedings of a seminar jointly organised by the Commission of the European Communities, Directorate-General for Science, Research and Development, and the Kernforschungsanlage Jülich GmbH, in the framework of the R & D programme on radioactive waste management and disposal, held in KFA Jülich, Federal Republic of Germany, 10–11 December 1985.

Chairmen: R. KRÖBEL (KFK)
H. DWORSCHAK (CEC)
E. MERZ (KFA)

Secretaries: L. CECILLE (CEC)
S. HALASZOVICH (KFA)

Radioactive Waste Management Series

DENITRATION OF RADIOACTIVE LIQUID WASTE

edited by

L. CECILLE

Commission of the European Communities,
Directorate-General for Science, Research and Development,
Nuclear Fuel Cycle Division

and

S. HALASZOVICH

Kernforschungsanlage Jülich GmbH

published by
Graham & Trotman
for the Commission of the European Communities

Published in 1986 by

Graham & Trotman Ltd　　**Graham & Trotman**
Sterling House　　　　　　**101 Philip Drive**
66 Wilton Road　　　　　　**Assinippi Park**
London SW1V 1DE　　　　　**Norwell, MA 02061**
UK　　　　　　　　　　　　**USA**

for the Commission of the European Communities,
Directorate-General Telecommunications, Information
Industries and Innovation, Luxembourg

EUR 10650

British Library Cataloguing in Publication Data

Denitration of radioactive liquid waste.
1. Radioactive waste disposal
2. Nitrification
I. Cecille, L.　II. Halaszovich, S.
III. Commission of the European Communities
621.48' 38　TD898

ISBN 978-0-86010-854-2　　　ISBN 978-94-011-9757-1 (eBook)
DOI 10.1007/978-94-011-9757-1

Introduction

The purpose of the Seminar was to provide an international interdisciplinary forum for in-depth discussions on the pros and cons of a denitration step by chemical means in the course of solidification of MAW (medium active waste) and HAW (highly active waste), prior to feeding the calciner or melter. In particular, differences in the volatility of some components and aerosol formation during the vitrification step are of interest. Chemical rather than thermal denitration may have particular merits which could simplify off-gas purification. A reduction in the salt burden of waste solutions, leading to a reduction in the volume of solid waste, may also be achieved.

The scientific programme of the Seminar consisted of invited lectures only, given by experts actively engaged in this field of research. The emphasis was on extensive discussions among the participants, with the aim of leading to an objective understanding of the state-of-the-art. The meeting also indentified issues and options for future research.

Summary

For some applications, such as concentration of high level liquid waste before interim storage, pretreatment of high level liquid waste prior to vitrification or actinide partitioning, and volume reduction of medium level liquid waste, the implementation of a denitration process by means of formic acid or formaldehyde seems very worthwhile. Moreover, for reducing technetium and ruthenium volatility during vitrification of high level waste, denitration with sugar proved to be quite efficient as far as the rotary kiln calciner is concerned. However, although some applications — such as concentration of high level liquid waste — have been operating on an industrial scale for many years, denitration depends on careful control of the waste composition as well as the operating conditions to avoid the occurrence of secondary interfering reactions which could be detrimental to the safety of the process or prevent the achievement of the desired final acidity and off-gas composition. Following a description of the basic reactions which feature denitration and the identification of the main parameters acting on reaction rate and induction time, the various applications of denitration reactions in nuclear waste management envisaged or currently practised are extensively analysed in the various presentations to the Seminar. Finally, conclusions and recommendations are included which highlight those aspects deserving further investigations.

CONTENTS

Session 3: Denitration of High Level Liquid Waste

OVERVIEW ON THE APPLICATION OF DENITRATION IN THE NUCLEAR FIELD

E.R. Merz
Institut für Chemische Technologie
der Nuklearen Entsorgung
Kernforschungsanlage Jülich, D-5170 Jülich, FRG

Summary

In recent years varying emphasis has been dedicated to the denitration of high active waste (HAW) and medium active waste (MAW) solutions generated in the nuclear fuel cycle, particularly in fuel reprocessing. These solutions contain free nitric acid and considerable amounts of nitrate salts. Denitration can be performed either thermally or in solution by chemical means. Various organic reductants such as formic acid, formaldehyde and sugar have been successfully applied. Yet another alternative is electrolytic denitration.
The main incentive of chemical denitration is reducing the acid content and in turn the salt load of the wastes formed by the neutralization of the acid with caustic soda. The ruthenium volatilization in the course of calcination and vitrification is also reduced. Further, the corrosion of waste storage tanks is minimized.
Conflicting opinions exist on whether or not the benefits arising from a chemical denitration of HAW concentrates are worthwhile the extra expenditures required. In case of MAW conditioning the situation is different if an incorporation into an organic matrix, e.g. bitumen or platics, is foreseen. Safety considerations may well justify a pre-treatment step.

1. INTRODUCTION

The nitrate system exhibits a widespread esteem in nuclear technology due to some of its advantageous properties. The most important of these are characterized by:

- The excellent dissolution power of nitric acid for actinide metals and their oxides, particularly thorium, uranium and plutonium;
- the oxidizing capability of nictric acid, especially at high concentrations;
- the nitrate-induced passivation of metallic construction materials and in turn their corrosion resistance;
- the good extraction behaviour of nitrate compounds with various organic solvents;
- the possibility of reconversion and recovery of nitrous oxides formed during chemical reactions;

- the ease of reconversion of nitrates into desirable forms,
  like oxides.
    The removal of excess nitric acid and nitrates from solutions to be
processed may be required in the course of chemical separation and waste
treatment procedures as well as in chemical conversion steps.
    Such treatment is commonly called denitration. Denitration may be
performed thermally by calcination and possibly by melting. A drawback of
this method is the formation of aerosols and possibly the oxidizing
environment which can promote an undesired volatilization of some elements into the off-gas.
    Better denitration can be achieved by chemical means in solution.
The major part of the nitric acid and nitrates will be destroyed, forming
N_2, N_2O, NO and NO_2, depending on the reducing agent and the conditions
of the chemical reaction. The nitrates are converted to oxides or oxide
hydrates. The most commonly used denitrating agents are formic acid,
formaldehyde and sugar.

2. INCENTIVES FOR CHEMICAL DENITRATION
2.1 Actinide nitrate solutions
    Most process flowsheets today in nuclear fuel production and repro-
cessing yield nitrate solutions of the actinides. These are ordinarily
converted into the oxide form in a series of process steps consisting of
concentration of the liquids, drying and calcination of the solids. The
excess nitric acid is removed by distillation or by chemical denitration.
A part of salt-bound nitrate can also be chemically destroyed. The
remainder has to be removed by thermal decomposition. A simple process
alternative includes evaporation of unbound water, removal of hydration
water and decomposition of nitrate salts. Dehydration and thermal denitration can occur simultaneously in the same reaction pot /1/.
    The chemical reactions that take place are different for the various
actinide compounds and for different methods of calcination. Prior to
calcination, some of the intermediate compounds, like pure uranyl nitrate
hexahydrate, melt and form a liquid. The further denitration is characterized by foaming of the syrupy concentrate and decrepiation of solid
particles /2/. Both effects are difficult to control and lead to severe
off-gas contamination, which is of particular importance in plutonium
handling.
    Better results are obtained by utilizing fluidized-bed calcination
/3/. However, a much smoother performance is observed by applying a chemical pre-denitration step, involving additional expenditure. Therefore,
there is still debate on whether or not one should follow this route.

2.2 HAW and MAW liquid effluents arising in fuel reprocessing
    The main sources of radioactive waste are fuel reprocessing plants.
More than 99 % of the total radioactivity generated in nuclear power production are contained in the aqueous raffinate stream of the first
solvent extraction cycle. This fraction is called high-level liquid waste
(HLLW). The remaining 1 % is distributed between roughly a one hundred
times bigger volume of medium and low level wastes of different types and
composition. From this one percent of radioactivity again at least 99.9 %
are attached to a so-called medium active fraction (MAW).

2

About half of the waste volumes arising in nuclear power production fall in a low activity category (LAW), but it contains not more than 1 part out of a million of the total radioactivity.

One of the main requirements concerning waste management is the minimization of waste volume to be disposed of in a repository. In order to achieve this goal, the salt content of liquid, high and medium level radioactive waste streams must be kept as low as possible. Since most of the HAW and MAW streams contain, besides the nitrate salts, variable concentrations of free nitric acid, their removal prior of solidification is of utmost importance in order to minimize the sodium nitrate formation by virtue of neutralization.

A former source of high nitrate salt loading of the HAW stream has been aluminium nitrate which was widely used as a salting-out reagent. Today its use has been abandoned in favor of pure nitric acid. Corrosion products play a minor role and make up to about 1 % of the total solid content in the waste solutions.

Neutralization of free nitric acid by adding of sodium hydroxide or carbonate always leads to an increased salt loading of the waste solutions and thus to undesired large waste volumes. In particular, MAW-streams stemming from solvent recovery and other plant rinsing and scrubbing operations produce high sodium nitrate containing waste solutions. The most effective countermeasure for reduced salt arisings is the avoidance of utilizing sodium hydroxide for neutralization of free nitric acid. This may be achieved in a threefold manner:

1. Removal of free acid by evaporation and distillation prior of neutralization.
2. Usage of ammonium hydroxide or carbonate instead of sodium hydroxide in order to form nitrate compounds which are easily decomposable by thermal treatments.
3. Denitration by chemical means applying suitable reducing agents. In this case also part of the salt-bound nitrate can be destroyed in addition to any free acid.

3. REMOVAL OF FREE NITRIC ACID BY EVAPORATION

Evaporation is a process whereby a solution is concentrated by vaporizing the solvent. Water or nitric acid are distilled off and the latter may be reused. Practically all the nonvolatile radioactive as well as inactive constituents remain at the evaporator bottoms.

Nitric acid and water form a maximum boiling azeotrope at 0.383 mole fraction of HNO_3. The boiling temperature is 121.9°C at 1 atm pressure. Figure 1 shows the boiling diagram.

The most concentrated acid that can be attained at the bottom of the fractionator is the azeotropic composition, whereas the distillate may approach pure water or pure nitric acid, depending upon whether the feed composition is, respectively, less than or greater than 0.383 mole fraction of nitric acid.

Due to the increasing solids content in the bottoms concentrate, a sludge will be formed after the solubility limits have been reached. A complete removal of free nitric acid is therefore hardly achievable.

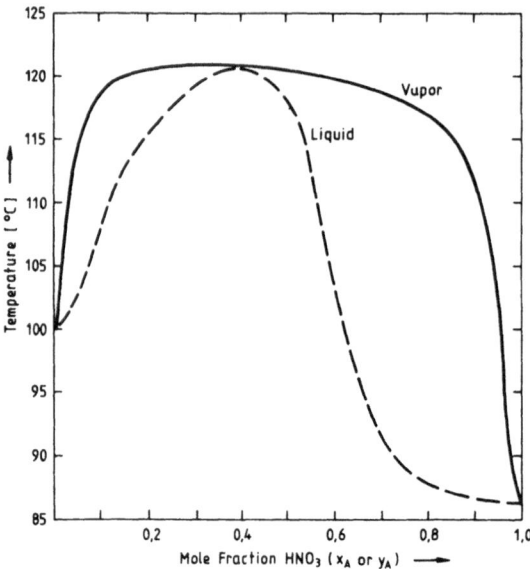

<u>Fig. 1:</u> Vapor-liquid relation for the system nitric acid-water at 1 atm. Temperature-composition diagram

Common problems in liquid waste evaporator operation are foaming, severe scaling and corrosion. To resist corrosion, evaporators are usually constructed of high quality stainless steel and operated at as low a temperature as is practicable. Scale has to be removed periodically, either mechanically or chemically. Foaming can possibly be avoided by foam-breaking devices inside the evaporator or by adding antifoam agents.

4. MERITS OF CHEMICAL DENITRATION OF LIQUID WASTE STREAMS

The usefulness of denitrating liquid waste streams has already been recognized in the late fifties /4/. At that time partial denitration of reprocessing liquid wastes stored in steel tanks has been proposed as a means of reducing corrosive attack and to minimize waste volumes. Instead of neutralizing excess nitric acid with sodium hydroxide it was proposed to destroy the nitric acid by addition of 1 mole of formaldehyde for each 3 moles of free acid. It was reported /4/ that the reaction proceeds smoothly and can easily be controlled.

As a consequence, chemical denitration applying formic acid, formaldehyde or sugar, has been recommended as a means of acid adjustment in waste storage and conditioning and for the sake of reducing materials corrosion. Important is the observation that although nitrate-induced passivation of several metals and metal alloys can be destroyed by adding formaldehyde and formic acid, these reagents do not depassivate stainless steel /5/.

The purpose for a chemical denitration in general is to suppress corrosion during tank storage by lowering the acidity close to a neutral

4

pH-value forming easily removable gaseous reaction products. Using this procedure instead of neutralizing the free acid minimizes the salt load of waste solutions. Research work indicated that by the combined action of hydrolysis and denitration, the denitrated solutions contain only nitrates associated with nonhydrolyzable alkali- and alkaline earth metals. Easily hydrolyzable cations, such as iron, zirconium, silver, etc. can be denitrated completely to oxides or hydroxides, whereas others like the ones from aluminium can only be destroyed with difficulties. A complete removal of nitrates from wastes with a high sodium content is impossible.

Another important advantage of a chemical denitration is the observed drastic decrease of ruthenium volatilization into the off-gas during HAW-vitrification. Ruthenium tetroxide (RuO_4) has been generally considered to be the volatile species responsible for the high carry-over into the off-gas system. Under the reducing conditions prevailing in chemical denitration, primary formed RuO_4 is reduced to nonvolatile RuO_2 /6/.

The dependence of ruthenium volatilization during glass melting as a function of nitrate concentration in the fission product calcinate is shown in figure 2. The ruthenium has been added as the nitrosyl ruthenium salt $RuNO(NO_3)_3$ spiked with Ru-106.

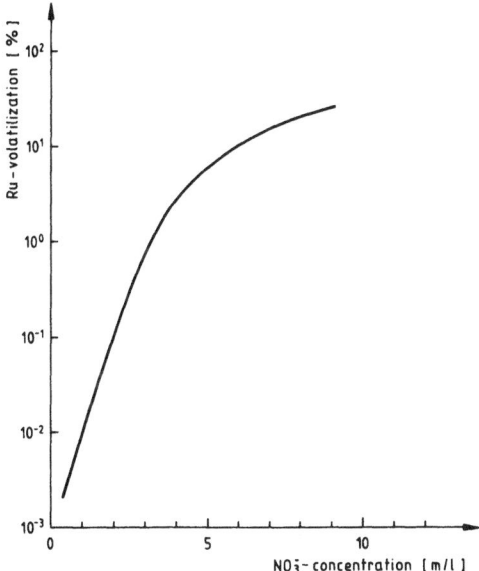

Fig. 2: Ruthenium volatilization as a function of the initial nitrate concentration in glass melt

The diagram clearly indicates that the volatility of ruthenium can effectively be reduced by chemical denitration of the liquid waste prior of vitrification. This has been confirmed by solidification comparison tests performed already in the early seventies at the Battelle Northwest Laboratories /7/. The ruthenium volatilization measured varied from 1 to 30 % depending on processing parameters.

Denitration of MAW-solutions prior of bituminisation or incorporation of their dried concentrates into a plastic matrix is of great importance in terms of safety, like avoidance of fire hazard due to uncontrolled exothermic reactions. Nitrates and hydrocarbons may violently react, leading to uncontrollable process conditions. The vigour of the reaction can even drastically be enhanced by catalytic action of certain trace constituents, like noble metals.

5. CHEMICAL VERSUS ELECTROLYTIC DENITRATION

For the sake of further reducing the final solid waste volumes to be disposed of, it has been proposed to apply an electrochemical denitration procedure which is characterized by a dispense with any additional redox chemicals needed for denitration /8,9/. Instead, electrolysis is applied which enables by adjustment of current density and electrode potential a controlled denitration reaction. Using a solid-bed cell, nitrates can be reduced alternatively to NO, N_2O or NH_3, depending on electrode potential and current yield:

$$HNO_3 + 3\ H^+ + 3\ e^- \rightarrow NO + 2\ H_2O$$
$$2\ HNO_3 + 8\ H^+ + 8\ e^- \rightarrow N_2O + 5\ H_2O$$
$$HNO_3 + 9\ H^+ + 9\ e^- \rightarrow NH_4^+ + 3\ H_2O$$

The competing reaction is a more or less pronounced hydrogen production. The most appealing advantage of an electrochemical denitration is its potential for a possible destruction of practically all salt-bound nitrates, even $NaNO_3$.
As suitable cathode materials platinum, titanium, silver, and graphit have been tested. As the anode, platinum or platinum-coated tantalum are used.
Although promising results have been reported by several authors, the picture is non-uniform and even contradictory /10/. This could be due to the catalytic action of some trace elements present in the electrolyte as well as on its reverse action of a strong electrode poisoning. As a consequence of pH rise in the electrolytic cell, precipitates of waste constituents are formed which can cause operational difficulties.
Not much is reported about the corrosion problems one may face with the construction materials of the cell body. Graphite and titanium are used in the test units.

6. CHEMICAL DENITRATION PROCEDURES
6.1 Formic acid
The feasibility of formic acid denitration has been demonstrated by several laboratories /11,12/. Two different techniques have been reported:

- adding of nitrate solution into excess of hot formic acid, or
- adding slowly concentrated formic acid in the desired quantity
  into boiling waste solution under a slow purge of inert gas.

6

A major concern of the former reaction, is loss of control over the reaction rate leading to under certain circumstances, formation of explosive formic acid-air mixtures. Also pricipitates consisting of formate salts may cause trouble. Therefore, it seems advisable to use the latter technique which avoids these drawbacks.

The principal reactions are /13/:

- if formic acid is in excess
$$2\ HNO_3 + 4\ HCOOH \rightarrow N_2O + 4\ CO_2 + 5\ H_2O$$
- if formic acid is limited and $HNO_3 > 4$ M
$$2\ HNO_3 + HCOOH \rightarrow 2\ NO_2 + CO_2 + H_2O$$
- if formic acid is limited and $HNO_3 < 2$ M
$$2\ HNO_3 + 3\ HCOOH \rightarrow 2\ NO + 3\ CO_2 + 4\ H_2O$$

However, the real behaviour is much more complex. The excess formic acid can be oxidized with hydrogen peroxide.

6.2 Formaldehyde

Formaldehyde is another suitable reagent for chemical denitration of liquid waste /14,15/. The reduction of HNO_3 by HCHO is illustrated by the two principal equations, the first is representative for the reaction at high acidity (8 - 16 M HNO_3) and the second at lower acidity (1 - 8 M HNO_3). They indicate that the addition of equimolar amounts of HCHO ensures the denitration of all HNO_3 independent of the acidity of the solution.

$$4\ HNO_3 + HCHO \rightarrow 4\ NO_2 + CO_2 + 3\ H_2O$$
$$4\ HNO_3 + 3\ HCHO \rightarrow 4\ NO + 3\ CO_2 + 4\ H_2O$$

Again two different techniques are feasible, either feeding formaldehyde into a hot waste solution or vice versa by adding the waste solution into diluted hot formaldehyde. The best results are obtained by feeding simultaneously both ingredients into the reaction vessel. In order to avoid the formation of burnable methyl nitrite (CH_3ONO), which is produced by a side reaction of methanol with nitrous oxide (the methanol is customary used as a stabilizing agent in formaldehyde) the use of paraformaldehyde or so-called process formaldehyde with less than 1 % methanol content is advisable.

No uncontrollable or dangerous process conditions have ever been observed using either one of the different process modes /16/.

An almost complete destruction and removal, respectively, of nitrate can be achieved if an equimolar amount of phosphoric acid is added to the waste solution in the course of denitration. However, the presence of phosphoric acid requires its incorporation into the final waste product, for example a phosphate glass may be produced.

6.3 Sugar

In the developing phase of waste management, the utilization of sugar has been proposed as a denitrating agent /17/. It has been observed that an improved decomposition of nitrate containing wastes with a high sodium content could be achieved during spray calcination of radioactive wastes /18/. It not only reduced the nitrate content in the product but also yielded a higher product density with free flowing properties. Some wastes, such as those with a high sulphate content, require the reducing action of sugar to calcine successfully.

It has recently become public that the French AVM process now also applies sugar as a curement aid in operating the rotary kiln.

The underlying reactions of nitrates with sugar are rather similar to the ones with formic acid and formaldehyde. The obvious differences are the amounts and kinds of nitrogenous oxides, nitrogen and carbon dioxide. The reaction with sugar is slower but on the other side does not require special controls.

$$C_{12}H_{22}O_{11} + 48\ HNO_3 \rightarrow 48\ NO_2 + 12\ CO_2 + 35\ H_2O$$

6.4 Combined action of hydrolysis and denitration

The degree of denitration achievable for MAW and HAW solutions depends on the content of free acid and hydrolyzable spezies.

The reaction proceeds stepwise. First the free nitric acid is destroyed. Then follows the hydrolyzation and successive denitration of nitrate salts according to the equations /19/:

first step
$$2\ Fe(NO_3)_3 + 9\ HCOOH \rightarrow 6\ NO + 9\ CO_2 + 6\ H_2O + 2\ Fe(OH)_3$$

second step
$$Al(NO_3)_3 + 6\ HCOOH \rightarrow N_2O + 4\ CO_2 + 5\ H_2O + Al(HCOO)_2NO_3$$

$$2\ Al(NO_3)_3 + 18\ HCOOH \rightarrow 3\ N_2O + 12\ CO_2 + 15\ H_2O + 2\ Al(HCOO)_3$$

Similar reaction schemes are valid for nitrates of rare earth elements, zirconium, etc..

7. CONCLUSIONS

Although chemical denitration of liquid waste exhibits advantages with regard to significantly lowering the quantity of waste solids to be stored as an alkaline salt cake, the main drawback of this approach is the enhanced corrosiveness of the solutions. High quality stainless steel is therefore required as a construction material.

The initial use of direct evaporation of HAW- and MAW-streams permits recovery of the bulk of the excess nitric acid. By such means the final nitric acid concentration can be reduced below 3 molar.

It has been extensively reported that the volatility of ruthenium is directly related to the acidity of the original liquor and hence the degree of nitration. Consequently, a number of studies have been made to tackle this problem. Addition of organic reductants, e.g. formic acid, formaldehyde or sugar will further reduce the ruthenium volatilization.

There are two effective and independent ways to suppress ruthenium volatilization in operating a liquid-fed ceramic glass melter. Off-gas cleanup does not seem to favor decisively the one or the other process scheme, since efficient scrubbing systems are readily at hand. The first one is to flood the molten glass pool more or less completely and continuously with liquid waste. The second one is denitration of the waste solution prior to feeding into the melter. Practical experience with the PAMELA plant at Mol/Belgium will reveal in time the most expedient procedure. Denitration could be helpful to keep the ruthenium

8

inventory in the recirculating liquid and gaseous process streams at the lowest level.

Electrochemical denitration, despite of its appealing features does not seem to be mature for technical application. Its future outcome seems doubtful.

For some MAW streams a pretreatment prior of incorporation into a suitable matrix, e.g. bitumen or plastics, is sometimes essential to destroy largely the nitrates, because their uncontrolled and vigorous reaction with organic matter could be detrimental to the health of personnel, process performance, equipment, and the quality of the resulting final solid waste product.

REFERENCES

1. GODBEE, H.W. and ROBERTS, J.T. (1961). Laboratory Development of a Pot Calcination Process for Converting Liquid Wastes to Solids. Report ORNL-2986
2. GEIER, R.G. (1957). Continuous Calcination Equipment for Converting UNH to UO_3. Report HW-49652 A, Hanford Atomic Product Operation
3. JONKE, A.A., PETKUS, E.J., LOEDING, J.W. and LAWROSKI, S. (1957). The Use of Fluidized Beds for Continuous Drying and Calcination of Dissolved Nitrate Salts. Nucl. Sci. Eng. 2, 203-319
4. EVANS, T.F. (1959). The Pilot Plant Denitration of PUREX Wastes with Formaldehyde. Report HAW-58587, Hanford Atomic Product Operation
5. ODOJ, R, MERZ, E. and WOLTERS, S. (1980). Effect of Denitration on Ruthenium Volatilization. Scientific Basis for Nuclear Waste Management, Vol. 2, Plenum Publishing Corporation, New York, N.Y./USA, 911-917
6. DUKE, E.M. (1960). Depassivation of 304L Stainless Steel in SULFEX Decladding Solutions, Report HW-65925, Hanford Atomic Product Operation
7. BLASEWITZ, A.G., RICHARDSON, G.L., McELORY, J.L., MENDEL, J.E. and SCHNEIDER, K.J. (1973). The High-Level Waste Solidification Demonstration Program. Proc. of a Symposium on the Management of Radioactive Wastes from Fuel Reprocessing, OECD-Paris, 27. Nov. - 1. Dec. 1972, 615-654
8. EPSTEIN, J.A. LEVIN, J. and RAVIN, S. (1964). Electrochemical Reduction of Dilute Nitric Acid. Electrochim. Acta 9, 1665-1673
9. BAUMGÄRTNER, F. and SCHMIEDER, H. (1978). Use of Electrochemical Processes in Aqueous Reprocessing of Nuclear Fuels. Radiochimica Acta 25, 191-210
10. KREYSA, G. and BREIDENBACH, G. (1980). Salpetersäure-Reduktion - eine elektrochemische Verfahrensstufe bei der Wiederaufarbeitung von Kernbrennstoffen. Chem.-Ing.-Techn. 52, 440-442
11. BRADLEY, R.F. and GOODLETT, C.B. (1972). Denitration of Nitric Acid Solutions by Formic Acid. Report DP-1299, Savannah River Laboratory
12. GUBER, W. et al. (1976). Lab.-scale and pilot-plant experiments on the solidification of high-level wastes at the Karlsruhe nuclear research center. Management of Radioactive Wastes from the Nuclear Fuel Cycle. Proc. IAEA Symp. Vienna, STI/PUB433, Vol. I, 271-281

13. HEALY, T.V. (1958). The Reaction of Nitric Acid with Formaldehyde and with Formic Acid and its Application to the Removal of Nitric Acid from Mixtures. J. Appl. Chem. 8, 553-561
14. FORSMAN, R.C. and OBERG, G.C. (1963). Formaldehyde Treatment of Purex Radioactive Wastes. Report HW-79622, Hanford Atomic Products Operation.
15. HALASZOVICH, St., DIX, S., HARMS, R. and SCHÄDLICH W. (1985). Auswahlkriterien für Denitrierungsverfahren in Brennelement-Wiederaufarbeitungsanlagen. Atomkernenergie/Kerntechnik 47, 94-96
16. DIX, S. (1985). FIPS II-Denitrierungsanlage - Aufbau und Betriebserfahrungen. Report JÜL-Spez-324, ISSN 0343-7639
17. BRAY, L.A. (1963). Denitration of Purex Wastes with Sugar. Report HW-76973, Rev. Hanford Atomic Products Operation
18. JOHNSON, B.M., Jr. (1961). Radiant Heat Spray Calcination. Report TID 7613, Bk. 1, 27-45
19. Orebough, E.G. (1976). Denitration of Savannah River Plant Waste Streams. Report DP-1417

CHEMICAL REACTIONS INVOLVED IN THE DENITRATION PROCESS WITH HCOOH AND HCHO

L. CECILLE, Commission of the European Communities,
(CEC) Brussels, Belgium
M. KELM, Kernforschungszentrum Karlsruhe GmbH, F.R.G.

Summary

Denitration reactions with HCOOH and HCHO proceed according to several reactions depending on the acidity and/or the redox potential of the reacting mixture. The parameters acting on the induction time - inherent to these reactions - and the denitration rate have been identified and their possible effects discussed. These mainly deal with the concentration of the reacting mixture components, the pressure in the reactor and the nitrous acid concentration. The basic outcome of this study is that gaseous compounds in equilibrium with nitrous acid catalyze the denitration reaction. As a consequence, every action which could remove these products from the reacting mixture extends the induction time and are detrimental to the denitration rate.

The conditions whereby the noble metals can affect the denitration process by modifying the reactions stoichiometry and by entailing oscillations of the reacting mixture are also presented and discussed. These phenomena were showed to occur in presence of rather high amounts of noble metals (those existing in HLLW) and iron. At last, based on the various experimental data observed, a tentative reaction mechanism is proposed involving a reaction between NO_2 on HCHO/HCOOH as an intermediate reaction between HCHO/HCOOH and HNO_3.

1. INTRODUCTION

Despite its industrial application to concentration of high level waste solutions for many years, the denitration process through the use of formic acid or formaldehyde has not been subject - except in F.R.G. - to a considerable number of investigations chiefly with respect to the fundamental chemical mechanisms involved. Even if the basic chemical reactions are well known, the identification of the various intermediate reactions participating in the overall process remains unclear.

The fact that the denitration of nitric acid proceeds with several reactions and that some chemical compounds are bound to catalyze the reaction render difficult the prediction of the reaction products evolving from the denitration of a particular waste stream. Accordingly unless the operating conditions are strictly kept constant, the straight-forward transposition of the results achieved for one specific denitration to another one is generally not possible. This explains why the various experimental denitration results quoted in literature are often different and in some cases somewhat contradictory.

Therefore the aim of this paper will be an attempt to clarify as far as possible the various reactions entering the denitration process with HCOOH or HCHO pointing out the parameters which are likely to influence the reactions and also to correlate, in the most logical way, the experimental observations performed here and there.

2. BASIC REACTIONS

Basically, a denitration process using HCOOH or HCHO consists of reducing nitric acid into volatile nitrogen and its oxides. However, due to the fact that nitrogen exhibits a large number of valency states, the reaction products may considerably vary according to the operating conditions chosen (oxidizing or reducing medium). Therefore, the reduction of nitric acid by formic acid or formaldehyde may proceed according to five main reactions.

2.1 Reactions with HCOOH

In strongly acidic solutions ($HNO_3 > 8M$), the reaction (1) was shown[1,2] to prevail whereas in more diluted HNO_3 solutions, mainly reactions (2) and (3) would occur:

$$2HNO_3 + HCOOH \longrightarrow 2NO_2 + CO_2 + 2H_2O \qquad (1) \quad HNO_3 > 8M$$

$$2HNO_3 + 2HCOOH \longrightarrow NO + NO_2 + 2CO_2 + 3H_2O \quad (2)$$
$$2HNO_3 + 3HCOOH \longrightarrow 2NO + 3CO_2 + 4H_2O \qquad (3)$$
$$0.5M < HNO_3 < 8M$$

When the denitration is performed in an excess of formic acid, (reducing medium), reactions (4) or (5) may be predominant depending on the concentrations ratio between formic acid and nitric acid[3,4].

$$2HNO_3 + 4HCOOH \longrightarrow N_2O + 4CO_2 + 5H_2O \qquad (4) \quad \text{moderate excess of HCOOH}$$

$$2HNO_3 + 5HCOOH \longrightarrow N_2 + 5CO_2 + 6H_2O \qquad (5) \quad \text{large excess of HCOOH}$$

2.2 Reactions with HCHO

Since oxidation of formaldehyde by nitric acid passes through generation of HCOOH as an intermediate reaction compound[2], the denitration reactions with HCHO are similar[1,5] to those identified with HCOOH with this difference that the consumption of nitric acid per mole of HCHO is bigger. Accordingly, as for HCOOH, there are five main reactions.

$$4HNO_3 + HCHO \longrightarrow 4NO_2 + CO_2 + 3H_2O \qquad (6) \quad HNO_3 > 8M$$

$$4HNO_3 + 2HCHO \longrightarrow 2NO_2 + 2NO + 2CO_2 + 4H_2O \quad (7)$$
$$4HNO_3 + 3HCHO \longrightarrow 4NO + 3CO_2 + 5H_2O \qquad (8)$$
$$0.5M < HNO_3 < 8M$$

$$4HNO_3 + 4HCHO \longrightarrow 2N_2O + 4CO_2 + 6H_2O \qquad (9) \quad \text{moderate excess of HCHO}$$

12

$$4HNO_3 + 5HCHO \longrightarrow 2N_2 + 5CO_2 + 7H_2O \qquad (10) \text{ large excess}$$
$$\text{of HCHO}$$

The similar behaviour of formic acid and formaldehyde during a denitration process is clearly demonstrated in Table 1 which shows that when HCHO and HCOOH are utilised in the same way, the resulting off-gas compositions are practically identical[6].

Table 1 : Balance of HNO_3 reduction into various volatile nitric oxides through denitration of HNO_3 + $NaNO_3$ mixture under different operating conditions.

nitrogen oxides / operating conditions	NO %	NO_2 %	N_2O %
HCHO feed into boiling HNO_3 + $NaNO_3$	71	5	24
HCOOH feed into boiling HNO_3 + $NaNO_3$	68	4	28
HNO_3 + $NaNO_3$ feed into boiling HCHO	28	2	70
HNO_3 + $NaNO_3$ feed into boiling HCOOH	22	3	75

However, due to the fact that generally formaldehyde is stabilised with 10-15%-vol. methanol, some variations between both organic reagents can be found as mentioned in Dr. Halaszovich's paper to this Seminar[3].

It must be stressed that the foregoing off-gas compositions correspond to an average of the compositions measured throughout a denitration experiment. Because, in a batch process, the respective concentrations of HCOOH/HCHO and HNO_3 are constantly varying in time, it is obvious that the off-gas composition as well as the stoichiometry of the reactions will be subject to changes. As an illustration of this, when a denitration is performed by feeding a nitric acid solution into concentrated and boiling formic acid, the stoichiometry of the reaction can vary in time from 2.25 to 1.85 as indicated in Fig. 1[7] demonstrating thus that the reactions (5), (4) and (3) successively influence or govern the overall denitration process.

3. INDUCTION TIME

The mechanism of denitration reactions is actually much more complex than the reactions (1) to (10) could suggest since an induction time is generally needed prior the reaction starts. From a pure safety standpoint, the existence of an induction time may be very detrimental to the process since large amounts of reaction products must not accumulate in the reactor in order to avoid any damageable overpressure at the starting-up of the reaction. Four main parameters can affect the duration of the induction time : the temperature, the respective concentrations of HCOOH/HCHO and HNO_3, the concentration of nitrous acid and very likely the nature of the surface state of the reactor walls.

3.1 Effect of the concentration of the reacting components and the temperature

The influence of the two first parameters has already been quantified by Holze[7] and Bradley and Goodlett[8] according to the empirical equations (11) and (12).

It is worth noting that the operating conditions used are quite different since the equation (11) has been determined with mixtures of HCOOH and HNO_3 concentrations varying from 2-12M and 0.2-0.4M respectively over a temperature range of 75-95°C whereas the equation (12) derives from experiments performed within concentration ranges of 1.75-6.75M and 1.1-11.4M for HCOOH and HNO_3 respectively for initial temperatures varying from 13°C to 50°C. Therefore, they are not straightforward comparable.

$$\frac{1}{t^*} = 2.8 \times 10^{29} \quad (HCOOH)^3 \quad (HNO_3) \quad \exp \left(- \frac{26040}{T} \right) \qquad (11)$$

$$\frac{1}{t^*} = 2.4 \times 10^{27} \quad (HCOOH)^3 \quad (HNO_3) \quad \exp \left(- \frac{14040}{T} \right) \qquad (12)$$

where t^* is expressed in min, T in °K and ( ) in M/l.

However, both equations (11) and (12) agree upon the fact that by operating a denitration at 100°C, the induction time is virtually negligible (a few seconds at maximum) but they are a bit conflicting about the part played by the HCOOH and HNO_3 concentrations on the duration of the induction time. Regarding this, some exploratory experiments have been carried out in KFA - Jülich by measuring the induction time for various HCOOH/HCHO and HNO_3 mixtures at room temperature (20°C). From these simple experiments, different lessons have been drawn. First, as far as HCOOH/HNO_3 mixtures are concerned, it was shown (see Table 2) that the higher the excess of HCOOH, the shorter the induction time confirming thus, qualitatively the merits of the equation (11). Surprisingly, when similar experiments were made with HCHO/HNO_3 mixtures, the exact opposite trend was observed[5].

Table 2: Induction time as a function of the HCOOH/HNO_3 molar ratio at room temperature

$\frac{HCOOH}{HNO_3}$	1.8	3.6	9.2
Induction time	10 min.	6 min. 50 s.	3 min. 50 s.

3.2 Effect of the reactor walls

These experiments also highlighted the unexpected importance of the walls of the reactor for the duration of the induction time since different induction times were noticed for denitration tests carried out with the same HCOOH/HNO$_3$ mixture but with various reactor configurations (see Table 3).

Table 3 : Effect of the surface (liquid/reactor) on the induction time
volume (liquid)
at room temperature (20°C) using (HCOOH)/(HNO$_3$) = 9.2

Surface (cm^2) Volume (cm^3)	3.2	3.0	1.4
Induction time	3 min. 50 s.	5 min. 20 s.	8 min. 20 s.

Accordingly, it might be possible that the surface state of the reactor walls influences the development of the intermediate liquid/gas reactions which are occuring and developing during the induction time.

3.3 Effect of the nitrous acid concentration

Besides the temperature, the concentration of HCHO/HCOOH and HNO$_3$ and secondarily of the reactor walls, the main important parameter governing the induction time is the concentration of nitrous acid in the reacting mixture. Actually, different authors[9] have found out that the induction time is strictly connected to the time needed for nitrous acid to be autocatalytically generated in the mixture to reach a threshold concentration of about 10^{-1} to 10^{-2} M/l. As a consequence, it is possible to shorten or even eliminate any induction time by simply adding some sodium nitrite to the mixture[10]. On the other hand, the presence of nitrous acid scavengers into the solution (e.g. sulfamic acid and hydrazine) may considerably increase the induction time[11].

3.4 Tentative explanation of the induction time mechanism

Whatever the organic reagent used HCOOH or HCHO, the induction time is characterized by some successive visual changes prior the reaction starts as showed in Fig. 2 to 5. First the mixture is turning yellow especially in the upper part of the reactor, then close to the liquid surface, a greenish zone is appearing. Once a dark green zone is formed, suddenly the temperature rises from 20°C to 95°C and the denitration starts with violency liberating red fumes. A rough explanation of the phenomena observed could be the following.

1. Generation of NO$_2$ along with HNO$_2$ (the mixture is turning yellow). Due to the slight solubility of NO$_2$, the yellowness is mainly developing in the upper part of the reactor.

2. Generation of N$_2$O$_3$ in the mixture according to the reaction (13) which is exothermic.

$$2HNO_2 \; \rightleftharpoons \; N_2O_3 + H_2O + Q \; cal \qquad\qquad (13)$$

3. Accumulation of N_2O_3 (blue compound) near the liquid surface which by superposition on the yellow colour (NO_2) gives rise to the green zone.

4. The heat developed by the reaction (13) accelerates the autocatalytical production of NO_2 and N_2O_3 which are released from the reactor (red fumes).

From these experiments, it results that gaseous compounds (e.g. NO_2) in equilibrium with nitrous acid should participate in the denitration mechanism, at least during the induction time.

4. DENITRATION RATE

The denitration rate depends on many parameters: the respective concentrations of HNO_3 and HCHO/HCOOH when mixing, the temperature, the nitrates concentration, the pressure over the reacting mixture as well as the presence in the waste solution of some chemical compounds capable to generate or destroy nitrites. This rather large number of parameters associated with the fact that different denitration reactions take place at the same time may explain the discrepancies between the various equations of denitration rate proposed in literature.

4.1 Influence of the concentration of the reacting components and the temperature

. Denitration rate with HCHO

J.B. Morris[12] determined different destruction rates of formaldehyde at boiling temperature according to the concentration of nitric acid in the mixture. When (HNO_3) > 1.4M, the destruction rate was shown independent on the nitric acid concentration according to equation (14), whereas at lower acidities, the kinetics law should obey to equation (15).

$$- \frac{d(HCHO)}{dt} = 2 \times 10^9 (HCHO)^8 \qquad\qquad (14)$$

$$- \frac{d(HCHO)}{dt} = 6 \times 10^{-3} (HNO_3) \quad with \; 4 \times 10^{-2} < (HCHO) < 3.6 \times 10^{-1} \; M/1 \qquad (15)$$

with t in min and (HCHO) and (HNO_3) in M/1.

The same author noticed that formic acid generated during reaction of formaldehyde with nitric acid also participated to denitration according to reaction (16) in the range 0.5M < HNO_3 < 4M with 0.1M < HCOOH < 0.7M.

$$- \frac{d(HCOOH)}{dt} = 0.2 \; (HCOOH)^2 (HNO_3)^{2.5} \qquad\qquad (16)$$

For his part, Healy[2] over a wide range of nitric acid concentrations (2M < HNO_3 < 15M) found out the reaction rate (17) which is quite

$$- \frac{d(HCHO)}{dt} = 0.028 \ (HCHO)^{1.4} \ (HNO_3)^3 \qquad (17)$$

different from that indicated by Morris.

. Denitration rate with HCOOH

According to Holze[7], within the concentration range 2M < HCOOH < 7M and 0.3M < HNO_3 < 1.5M where reaction (18) prevails, the denitration rate could be expressed by :

$$- \frac{d(HNO_3)}{dt} = k \ (HCOOH)^2 \ (HNO_3)^3 \qquad (18)$$

where k = 0.07 $(1/M)^4$ min^{-1} at 100°C

It is worth noting that variation of k as a function of temperature follows the Arrhenius law : $k = k_o \ e - A/T$

with k_o = 1.45 x 10^{16} $(1/M)^4$ s^{-1} and A = 16406 K

By operating in an excess of nitric acid (2M < HNO_3 < 15M), Healy determined a third order reaction over the nitric acid at constant formic acid concentration. In comparison with similar experiments performed with formaldehyde, denitration with formic acid appeared to double the reaction rate.

More recently, Saum[13] determined other rate equations in case of denitration of simulated HLLW by means of formic acid. For nitric acid concentrations varying within the range 1M < HNO_3 < 5M, he quoted the equation (19) whereas in a lower range of acidity 0.4M < HNO_3 < 1.7M

$$- \frac{d(HNO_3)}{dt} = 0.0354 \ (HNO_3)^{1.6} \ (HCOOH)^{1.5} \qquad (19)$$

a somewhat different rate equation (20) was determined:

$$- \frac{d(HNO_3)}{dt} = 0.021 \ (HNO_3)^{1.4} \ (HCOOH)^{1.3} \qquad (20)$$

The review of all these rate equations highlights the huge discrepancies existing between them especially concerning the partial reaction orders. Even if only rate equations dealing with denitration with HCOOH in a nitric acid medium are considered, hence the equations (18), (19) and (20), the discrepancies remain important suggesting that other parameters should intervene in the denitration process.

In addition, the fact that, in some cases the rate equations exhibit an unusually high reaction order (i.e. 3 for the equation (18)) confirms that the decomposition of HNO_3 with HCOOH follows a complex reaction mechanism.

4.2 Influence of the salt content

Healy[2] reported that nitrates favour the denitration rate and decrease the induction time. He attributed this effect to the formation of undissociated HNO_3 species which should play an important part in the reaction mechanism.[3] Such an effect can easily be demonstrated when denitrating a nitric acid solution with/without sodium nitrate by an excess of formic acid (see Table 4).

In absence of sodium nitrate, even if 80% of the nitric acid are destroyed within the first half hour after the addition of formic acid, the destruction of the remaining 20% is very low, whereas with sodium nitrate, the reaction is completed after only two hours boiling under reflux. Although the effect of sodium nitrate on the denitration rate is obvious, the stoichiometry of the reaction does not much vary under these conditions.

The effect of nitrates on the denitration rate may also explain the variations of efficiency quoted by Kelm et al[14] during denitration experiments on MLLW containing variable amounts of sodium nitrate.

Table 4 : Effect of sodium nitrate on the denitration rate and stoichiometry

(addition of conc. HCOOH into boiling nitric acid solutions)

Operating conditions	Consumption of (HCOOH) with respect to (HNO_3)	(HNO_3) after 2 h. boiling under reflux	pH after 22 h. boiling under reflux
4.1 M HNO_3 + 8.2 M HCOOH	1.6	0.20 M	1.1
4.1 M HNO_3 + 2.0 M $NaNO_3$ + 8.2 M HCOOH	1.5	n.d.	2.09

n.d. = not detectable

4.3 Effect of hydrochloric acid (protonic effect)

Holze[7], who performed extensive investigations on denitration with HCOOH, also studied the possible effect of hydrochloric acid on the denitration rate. Substantially, he demonstrated that the presence of HCl along with HNO_3 only favours the denitration rate since in this case, the reaction constant k increases by a factor of eight (see Fig. 6). He also determined the reaction rate when the denitration is performed in an excess of formic acid.

$$- \frac{d(HNO_3)}{dt} = k \ (HNO_3)^2 \ (HCOOH)^2 \ (HCl) \qquad (21)$$

with $k = 1.82 \times 10^{20} \exp\left(-\frac{19185}{T}\right) \quad (1/M)^4 \ s^{-1}$

4.4 Influence of the pressure

The pressure over the reacting mixture was proved[15] to be an important parameter which may influence much the denitration rate. This has been demonstrated for the specific purpose of HLLW denitration with formic acid within a rather narrow range of pressure (- 270 mm H_2O until + 300 mm H_2O). The procedure adopted for these experiments consisted of feeding simulated HLLW (200 ml/h) into boiling and concentrated HCOOH using an initial $HCOOH/HNO_3$ molar ratio equal to 2.

The destruction kinetics of nitric acid thus determined (see Fig.7) show that under these operating conditions relatively small variations of pressure may strongly affect the denitration rate. This experimental observation confirms that at least some of the intermediate chemical species which catalyze the denitration reaction are gaseous.

This statement is in line with the observations previously made with respect to the induction time suggesting that NO_2 is a fundamental intermediate compound of the reaction.

It must be pointed out that the sensitivity of the denitration to pressure was mainly recorded with low feeding rate. It is expected that using higher feeding rate this influence should decrease

Consequently, in this case it could appear helpful to perform denitration under a slight overpressure (+ 100 mm H_2O) in order to achieve faster reaction rates as it was practised at the CEA-Fontenay-aux-Roses[16].

To a certain extent, the sensitivity of denitration reactions to pressure could explain why for similar experiments some discrepancy appear in the literature concerning equation rates.

It must be pointed out that the influence of such a parameter on the denitration rate can be counterbalanced by acting on the other parameters which favour the reaction. Thus, it has been experimentally proven that in adopting a high $HCOOH/HNO_3$ molar ratio for denitrating simulated HLLW (2.5 instead of 2) the part played by pressure variations on the overall denitration rate can become marginal. In the same way, an increase of the nitrate concentration may have a comparable positive effect as well as an increase of the feeding rate. To be comprehensive on this item, it must be said that the pressure effect on the denitration rate was proved to be less pronounced as far as diluted $HCOOH/HNO_3$ mixtures are concerned (0.5 M HNO_3, 1 M HCOOH) as demonstrated by Holze[7].

4.5 Influence of air

Since the oxygen of air may intervene in the denitration reaction e.g. by oxidizing NO into NO_2, it was considered worthwhile to study its possible effect. To this end similar denitration experiments with formic acid were carried out alternatively in presence of air or argon[15]. Since no significant variations of the HNO_3 and HCOOH destruction kinetics were found out, it is concluded that the oxygen does not participate in the denitration mechanism.

However, an experimental observation deserves to be mentioned which deals with release of NO_2 (red fumes) in the reactor at the beginning of the reaction when feeding a nitric solution into boiling and concentrated formic acid even in presence of argon which is contradictory with the reactions (4) and (5). This is another indication that very likely NO_2 is an intermediate reaction compound.

4.6 Bubbling effects

The effect of bubbling air or argon under the surface of a mixture of HCOOH and HNO_3 to favour mixing was shown to slow down the denitration rate [15]. Most probably, this has to be attributed to removal of gaseous species in equilibrium with the autocatalytic agents. These results confirmed those determined in the past by Bradley and Goodlett [8].

5. EFFECT OF THE NOBLE METALS

Despite qualitative evidence that some noble metals (Pd and Rh) can accelerate the denitration between formic acid and nitric acid [17], Saum [13] demonstrated that such an effect, if any, would be marginal. Actually, the action of noble metals, on the one hand, may affect the HCHO/HCOOH consumption and on the other hand can entail the occurrence of oscillations in the reacting mixture by means of secondary reactions [7].

5.1 Overconsumption of HCHO/HCOOH

When the waste stream to be denitrated contains high amounts of noble metals (e.g. concentrated HLLW), an overconsumption of HCOOH has been noticed especially at low nitric acidity [16]. This overconsumption can result from the reactions (22) and (23) which successively take place.

$$Pd^{2+} + HCOOH \longrightarrow 2H^+ + CO_2 + Pd \text{ (black)} \qquad (22)$$

$$HCOOH \xrightarrow{\text{Pd (black)}} H_2 + CO_2 \qquad (23)$$

Accordingly, the presence of noble metals enables denitration of HLLW to reach neutrality (pH 7-8) simply by implementing the reactions (22) and (23) by keeping the denitrated waste solution boiling under reflux for a few hours.

However, it must be pointed out that the release of H_2, up to 10% - vol according to Kelm et al [14], in the off-gas may need some dilution with an inert gas in order to comply with safety requirements.

Another possible consequence of the generation of hydrogen during denitration is that some elements or chemical compounds may be reduced as it was envisaged for Pu [16], Fe [17] and as it also results from the release of ammonia through the reaction [18] (24)

$$2NO + 5H_2 \xrightarrow{\text{(Pd catalyst)}} 2NH_3 + 2H_2O \qquad (24)$$

5.2 Occurrence of oscillations

Another possible effect of noble metals on denitration deals with the occurrence of oscillations in the reacting mixture, hence in the off-gas production (see Figure 10) especially when Fe and Pd are present [7].

It is worth noting that such a phenomenon is recorded when a nitric solution is slowly fed into boiling HCOOH especially towards the end of the feeding phase.

If it is assumed that the critical parameter governing the denitration rate is the concentration of nitrous acid in the reacting mixture, the occurrence of oscillations could result from its periodical consumption by some element or chemical species present in the solution. Therefore a possible explanation of the observed phenomena could be the following:

1. No oscillations occur as long as elemental palladium is not precipitated. This is always observed during the first part of the feeding period. At a high dosage rate the nitric acid concentration in the reacting mixture is kept high enough to avoid precipitation of Pd over the whole feeding phase and thus the occurrence of oscillations.
2. At lower feeding rates, the nitric acid concentration can drop below a critical value during the feeding phase so that palladium-black is formed according to reaction (22).
3. The presence in the reacting mixture of palladium black, ferric nitrate and an excess of formic acid causes the HNO_2 consumption – hence a slow -down of the denitration rate – according to the reaction ^{19} (27) resulting from various intermediate reactions.

$$3HCOOH \xrightarrow{\text{(Pd black)}} 3 H_2 + 3 CO_2 \qquad (23)$$

$$1/2 H_2 + Fe^{3+} \longrightarrow Fe^{2+} + H^+ \qquad (25)$$

$$Fe^{2+} + H^+ + HNO_2 \longrightarrow Fe^{3+} + NO + H_2O \qquad (26)$$

$$NO + 5/2 H_2 \longrightarrow NH_3 + H_2O \qquad (24)$$

$$HNO_2 + 3HCOOH \longrightarrow NH_3 + 3 CO_2 + 2 H_2O \qquad (27)$$

4. Because of continuous feeding, progressivily the nitric acid concentration increases in the reacting mixture. After reaching a certain concentration level, palladium black is redissolved through the reaction (28) and a new cycle starts.

$$3 Pd + 2HNO_3 + 6 H^+ \longrightarrow 3 Pd^{2+} + 2NO + 4H_2O \qquad (28)$$

6. PROPOSAL FOR A DENITRATION MECHANISM

On the basis of the foregoing results, experimental observations, and literature data, i.e.:
  - autocatalytically generation of HNO_2 during the induction period
  - release of NO_2 from the reacting mixture even in strongly reducing medium
  - effect of pressure on the denitration rate.
It is possible to propose which main intermediate reactions could intervene in the overall basic reactions (1) to (10).

6.1 Reactions with HCOOH in strongly acidic solutions

$$2HNO_3 + 2HNO_2 \rightleftharpoons 4NO_2 + 2H_2O \quad \text{(slow reaction)} \quad (29)$$

$$2NO_2 + HCOOH \longrightarrow 2HNO_2 + CO_2 \quad (30)$$

$$2HNO_3 + HCOOH \longrightarrow 2NO_2 + 2H_2O + CO_2 \quad (1)$$

6.2 Reactions with HCOOH in moderately acidic solutions

$$6HNO_3 + 6HNO_2 \rightleftharpoons 12NO_2 + 6H_2O \quad \text{(slow reaction)} \quad (29)$$

$$6NO_2 + 3HCOOH \longrightarrow 6HNO_2 + 3CO_2 \quad (30)$$

$$6NO_2 + 2H_2O \rightleftharpoons 2NO + 4HNO_3 \quad (31)$$

$$2HNO_3 + 3HCOOH \longrightarrow 4H_2O + 3CO_2 + 2NO \quad (3)$$

6.3 Reactions with HCOOH in an excess of formic acid

$$8HNO_3 + 8HNO_2 \rightleftharpoons 16NO_2 + 8H_2O \quad \text{(slow reaction)} \quad (29)$$

$$8NO_2 + 4HCOOH \longrightarrow 8HNO_2 + 4CO_2 \quad (30)$$

$$9NO_2 + 3H_2O \rightleftharpoons 3NO + 6HNO_3 \quad (31)$$

$$3NO \longrightarrow N_2O + NO_2 \quad \text{(reaction} \quad (32)$$
possible in a re-[18]
ducing medium)

$$2HNO_3 + 4HCOOH \longrightarrow N_2O + 4CO_2 + 5H_2O \quad (4)$$

6.4 Reactions with HCHO in strongly acidic solutions

$$2HNO_2 + 2HNO_3 \rightleftharpoons 4NO_2 + 2H_2O \quad \text{(slow reaction)} \quad (29)$$

$$NO_2 + HCHO \longrightarrow HCOOH + NO \quad (33)$$

$$2NO_2 + HCOOH \longrightarrow 2HNO_2 + CO_2 \quad (30)$$

$$NO + 2HNO_3 \rightleftharpoons 3NO_2 + H_2O \quad \text{(shifted} \quad (31)$$
towards the right in
concentrated nitric
acid)

$$4HNO_3 + HCHO \longrightarrow 4NO_2 + 3H_2O + CO_2 \quad (6)$$

6.5 Reactions with HCHO in moderately acidic solutions

$$6HNO_2 + 6HNO_3 \rightleftharpoons 12NO_2 + 6H_2O \quad \text{(slow reaction)} \quad (29)$$

$$3NO_2 + 3HCHO \longrightarrow 3HCOOH + 3NO \quad (33)$$

$$6NO_2 + 3HCOOH \longrightarrow 6HNO_2 + 3CO_2 \quad (30)$$

$$3NO + H_2O \rightleftharpoons NO + 2HNO_3 \quad \text{(shifted} \quad (31)$$
towards the right in concentrated nitric acid)

$$4HNO_3 + 3HCHO \longrightarrow 4NO + 5H_2O + 3CO_2 \quad (8)$$

6.6 Reactions with HCHO in an excess of formaldehyde

$$4HNO_2 + 4HNO_3 \rightleftharpoons 8NO_2 + 4H_2O \quad \text{(slow reaction)} \quad (29)$$

$$2NO_2 + 2HCHO \longrightarrow 2HCOOH + 2NO \quad (33)$$

$$2HCOOH + 4NO_2 \longrightarrow 4HNO_2 + 2CO_2 \quad (30)$$

$$3NO_2 + H_2O \rightleftharpoons NO + 2HNO_3 \quad (31)$$

$$3NO \longrightarrow N_2O + NO_2 \quad \text{(reaction} \quad (32)$$
possible in a reducing medium [18])

$$2HNO_3 + 2HCHO \longrightarrow N_2O + 3H_2O + 2CO_2 \quad (9)$$

Similar sets of intermediate reactions can be proposed for the basic reactions (2), (5), (7) and (10).

According to these reactions, it can be understood why the nitrates favour the denitration rate because they promote the reaction between HNO_3 and HNO_2 which is normally slow. The blocking-up action of anti-nitrites on the denitration rate can also be easily explained by preventing generation of NO_2.

With respect to the important role of nitrous acid during the induction period and its permanent regeneration in course of the denitration reaction, this can be explained from the following reactions which involve an autocatalytical increase of the HNO_2 concentration.

$$2HNO_3 + 2HNO_2 \rightleftharpoons 4NO_2 + 2H_2O \quad (29) \text{ Basic equilibrated}$$
reaction in a nitric acid medium.

$$4NO_2 + 2HCOOH \longrightarrow 4HNO_2 + 2CO_2 \quad (30) \text{ Doubling of the } HNO_2$$
concentration which shifts the reaction (29) towards the right.

7. CONCLUSIONS

The reactions between HCHO or HCOOH with HNO_3 during a denitration process are numerous and depend on the acidity and/or the redox potential of the reacting mixture. Accordingly the procedure adopted for denitrating a given liquid waste (feeding into an excess of HCHO/HCOOH or inversely) governs the dominating reactions as well as the off-gas composition.

The induction time which is inherent to denitration reaction can be reduced to a few seconds by addition of some sodium nitrite and by operating at boiling temperature. It is of paramount importance to make sure that no anti-nitrites compounds are present in the waste solution.

Amongst the various parameters which favour the denitration rate, the most important ones are the concentration of the reacting components (HCHO/HCOOH and mainly HNO_3), hence the feeding rate, the temperature, the presence of nitrates and in certain cases a slight overpressure in the reactor.

The presence of noble metals in the waste solution is proved to affect the denitration through secondary reactions entailing a further consumption of HCHO/HCOOH. Under certain operating conditions, these secondary reactions may give rise to H_2 generation and to oscillations in the reacting mixture.

Experimental observations (colour changes and gas development during the induction time and influence of a slight overpressure or gas bubbling on the denitration rate) and the need for HNO_2 to reach a concentration threshold prior the reaction starts can be correlated with each other if it is supposed that the HNO_2/NO_2 couple are the fundamental intermediate compounds which participate in the denitration mechanism.

Therefore, all the chemical species (e.g. Fe^{2+} associated with low feeding rate) which can sufficiently lower the nitrous acid concentration in the reacting mixture and some particular operating conditions (e.g. slight depressure combined with low feeding rate) which could remove NO_2 from the solution are detrimental to the denitration rate. Consequently the prediction of the evolution of a denitration reaction applied to a specific stream does not seem possible without a thorough chemical analysis of its main components and a strict control of the operating conditions.

To make sure that, in any time, a denitration proceeds steadily, the setting-up of a mean to in-situ control of the evolution of the reaction would be worthwhile and recommended.

REFERENCES

1. T.V. Healy and B.L. Ford
   "The destruction of nitric acid by formaldehyde, part I"
   AERE C/R 1339 (April 1954)

2. T.V. Healy and B.L. Davies
   "The destruction of nitric acid by formaldehyde, parts II, III and IV"
   AERE C/R 1739 (February 1956)

3. G. Koch, Z. Kolarik, H. Haug, W. Hild and S. Drobnik
   "Proceedings management of radioactive wastes from fuel reprocessing"
   OECD Paris (1972), p. 1081

4. K. Holze, H.D. Finke, M. Kelm and W.D. Deckwer
   "Reaction model for denitration with formic acid of waste effluents from nuclear fuel reprocessing plants"
   Ger. Chem. Eng. 2 (1979) 361-371

5. S. Halaszovich, S. Dix and R. Harms
   "Denitration of reprocessing concentrate by means of HCHO"
   Paper presented at this Seminar

6. S. Halaszovich, S. Dix and R. Harms
   Final report for the C.E.C. Contract nr. 178-81-31 and 286-82-31 WASD

7. K. Holze
   "Dissertation"
   Univ. Hannover (December 1980)

8. R.F. Bradley and C.B. Goodlett
   "Denitration of nitric acid solutions by formic acid"
   DP 1299 (June 1972)

9. J.V.L. Longstaff and K. Singer
   "The kinetics of oxidation by nitrous acid and nitric acid - Parts I and II"
   J.C.S. (1954) p. 2604-2617

10. M. Kubota, I. Yamaguchi and H. Nakamura
    "Effects of nitrite on denitration of nuclear fuel reprocessing waste with organic reductants"
    J. Nucl. Sci. Technol. Vol. 16, No 6 (June 1979)

11. M. Germain, A. Bathellier and P. Bérard
    "Use of formic acid for the stripping of plutonium"
    Proceedings of the International Solvent Extraction Conference (Lyon 1974) p. 2075-2087

12. J.B. Morris
    "La réaction de l'acide nitrique sur le formaldehyde"
    Energie nucléaire, n° 1, p. 216-224 (1957)

13. C.J. Saum, L.H. Ford and N. Blatts
    "The denitration of simulated fast reactor highly active liquor waste"
    DCO 7387 (S), Proceedings of the international Seminar on chemistry and process engineering for high level liquid waste solidification – KFA Jülich (1-5 June 1981)

14. M. Kelm, B. Oser, S. Drobnik and W.D. Deckwer
    "Denitration of aqueous waste solutions from the nuclear fuel reprocessing"
    Nuclear Technology, vol. 51 Nov. 1980, p. 27-32

15. L. Cécille and G. Tanet
    "Etude de l'influence de quelques paramètres sur la réaction de dénitration d'une solution HLLW simulée – Schéma de séparation des actinides HDEHP"
    Note technique n° 1.07.03.80.98 – JRC Ispra (Nov. 1980)

16. L. Cécille and M. Lecomte
    "Denitration of HLLW for actinide partitioning"
    Paper presented at this Seminar.

17. H. Krause et al.
    "Annual report for 1972" KfK-2000

18. P. Pascal
    "Traité de Chimie Minérale"
    Ed. Masson

19. E. Abel, H. Schmid and F. Pollak
    "Kinetik der Oxydation von Ferro-Ion durch Salpetrige Säure"
    Monatsherfte für Chemie Wien (1936) $\underline{69}$, 125

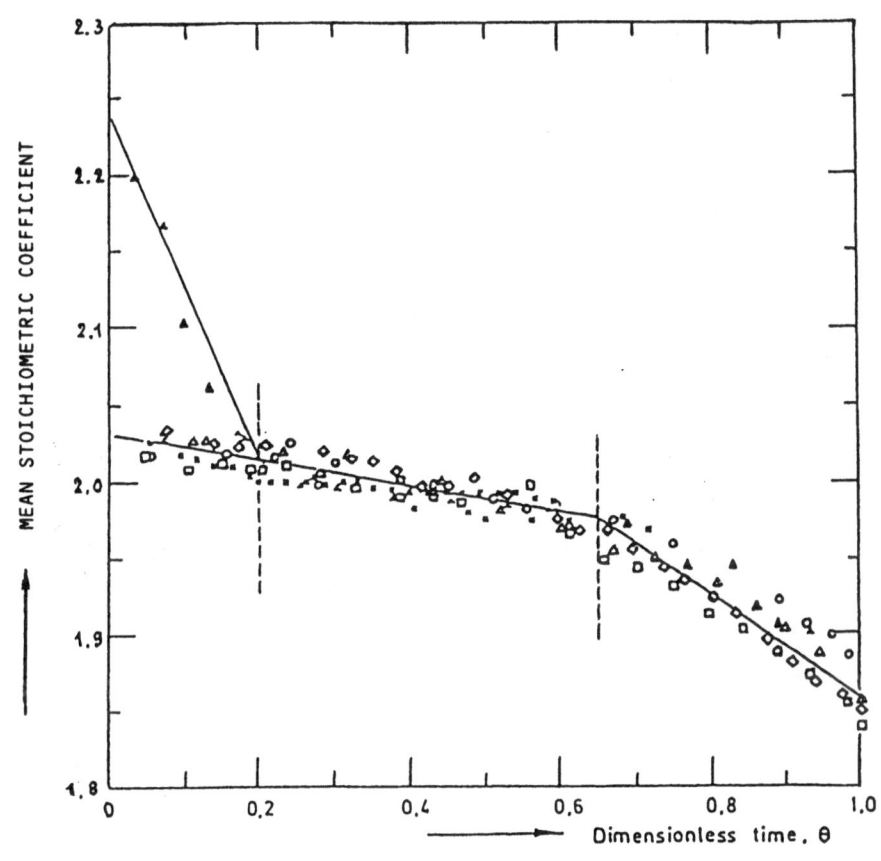

FIG. 1   VARIATION OF THE STOICHIOMETRIC COEFFICIENT (MOLE HCOOH/MOLE HNO$_3$) AS A FUNCTION OF THE FEEDING TIME[7]

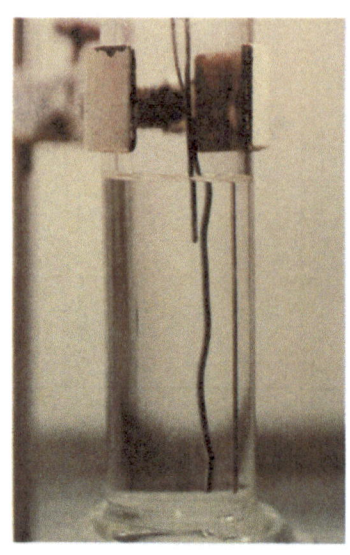

Fig. 2   t = o

The mixture is colourless (T=20°C)

Fig. 3   t = 1 min.

The mixture is turning yellow (T=20°(

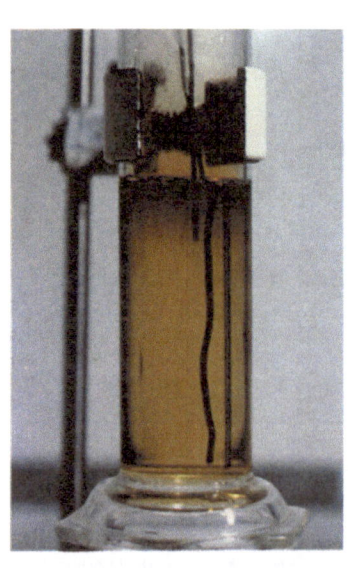

Fig. 4   t = 3 min.

A greenish zone is developing near
the liquid surface   (T = 20°C)

Fig. 5   t = 3 min. 50 s.

The denitration reaction starts in
the green zone   (T = 95°C)

EVOLUTION OF A HCOOH/HNO$_3$ MIXTURE AT ROOM TEMPERATURE DURING THE INDUCTION TIME

$\frac{(HCOOH)}{(HNO_3)}$ = 9.2   (HNO$_3$) = 14.4 N   Total volume : 60 ml

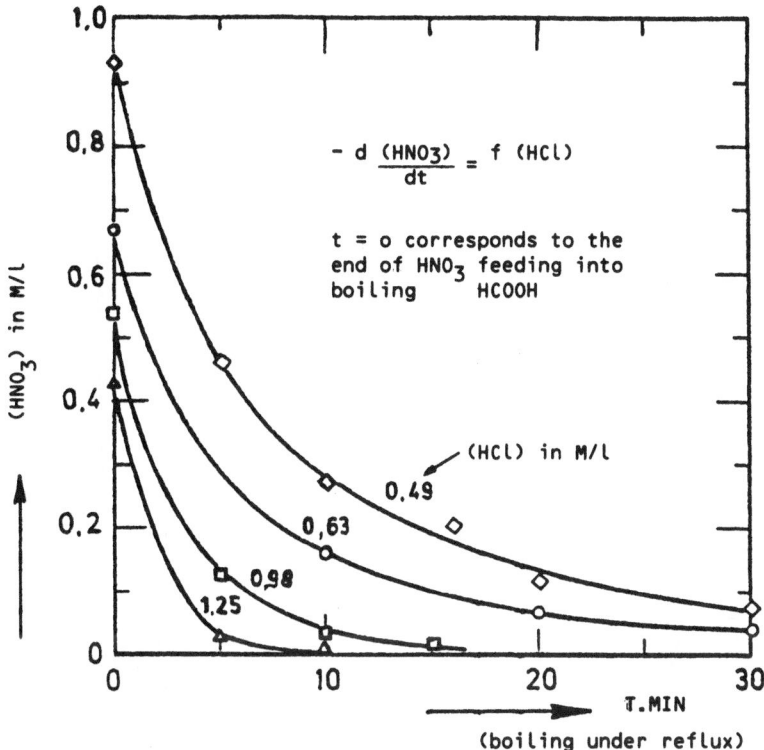

The figure shows a plot with:
- Y-axis: (HNO_3) in M/l, ranging from 0 to 1,0
- X-axis: T.MIN (boiling under reflux), ranging from 0 to 30

Text within the plot:

$$- d \frac{(HNO_3)}{dt} = f (HCl)$$

t = o corresponds to the end of HNO_3 feeding into boiling   HCOOH

(HCl) in M/l

0,49
0,63
0,98
1,25

FIG. 6   INFLUENCE OF HCL ON THE NITRIC ACID DESTRUCTION RATE[7]

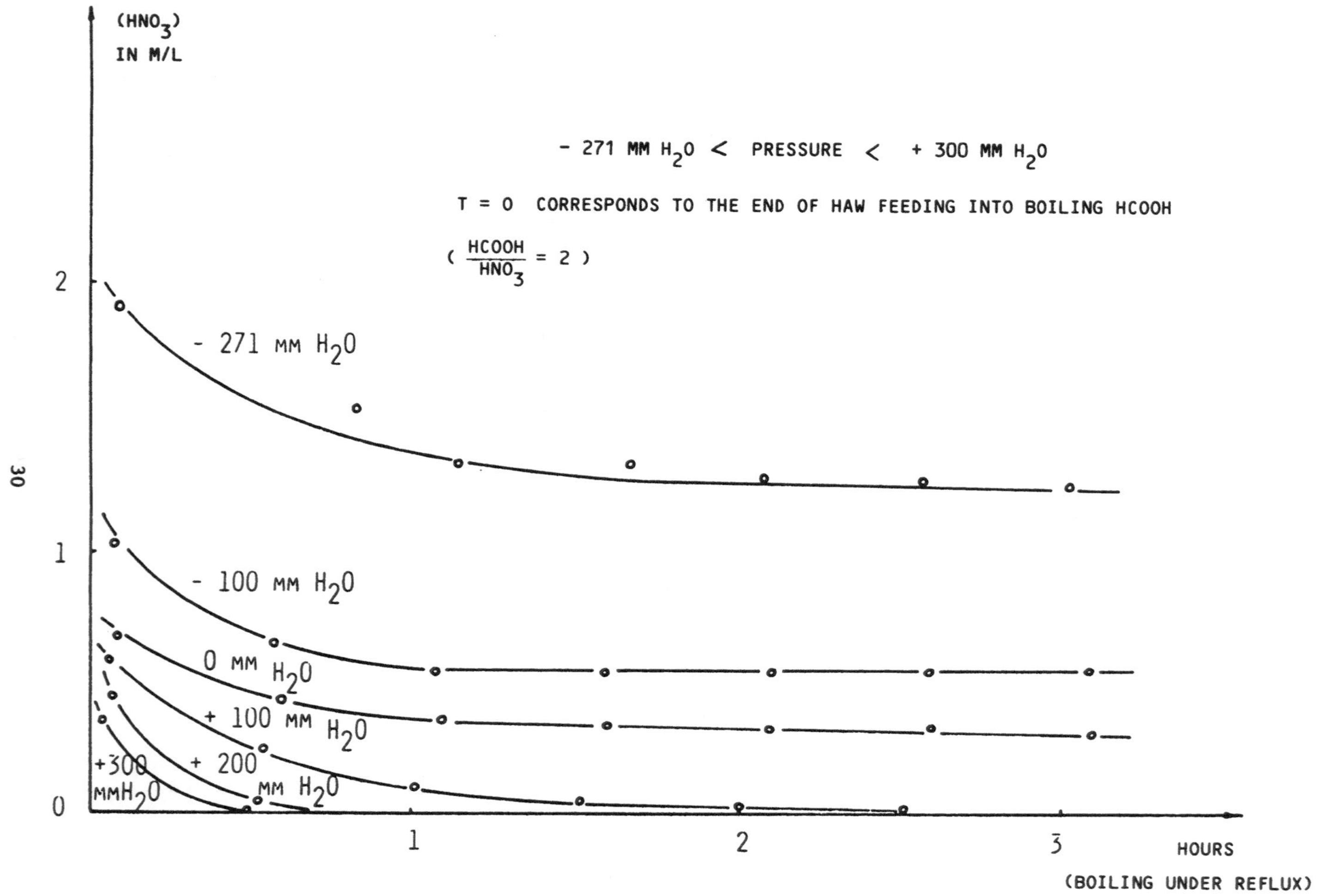

- 271 MM H₂O < PRESSURE < + 300 MM H₂O

T = 0 CORRESPONDS TO THE END OF HAW FEEDING INTO BOILING HCOOH

($\frac{HCOOH}{HNO_3}$ = 2)

(HNO₃) IN M/L

- 271 MM H₂O

- 100 MM H₂O

0 MM H₂O

+ 100 MM H₂O

+ 200 MM H₂O

+ 300 MM H₂O

HOURS

(BOILING UNDER REFLUX)

FIG. 7   DESTRUCTION KINETICS OF HNO₃ FOR RELATIVE PRESSURES

30

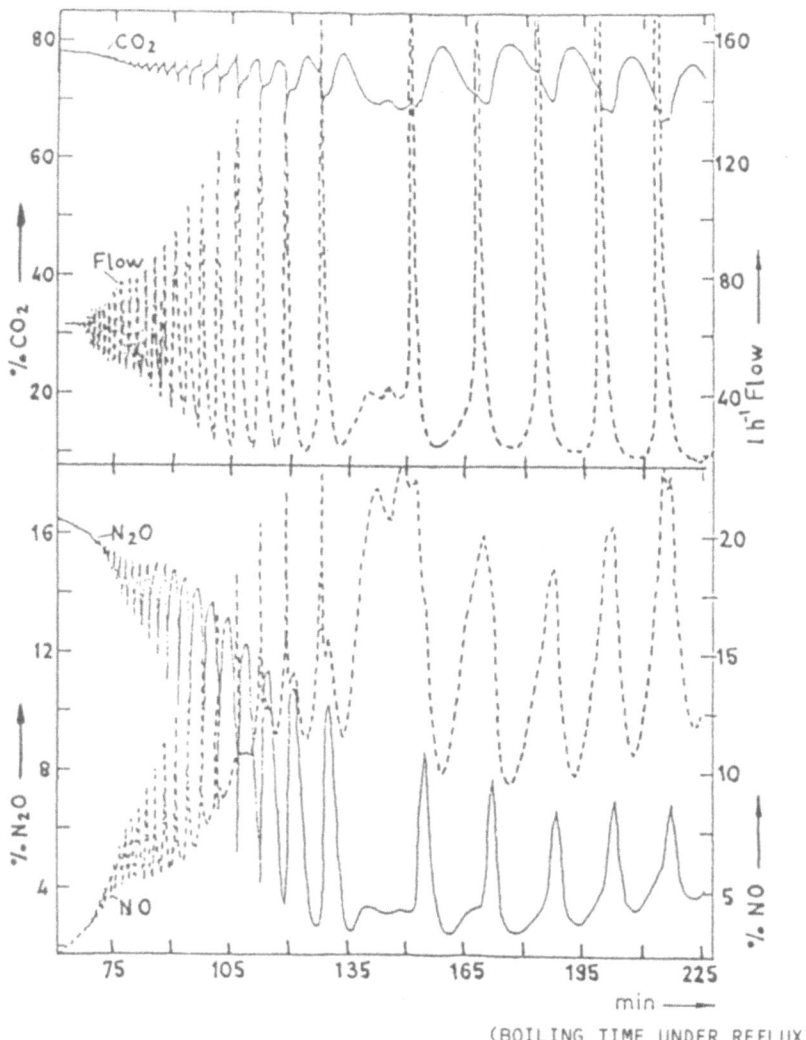

(BOILING TIME UNDER REFLUX)

FIG. 8   OSCILLATIONS IN THE OFF-GAS EVOLUTION DURING DENITRATION OF
Fe/Pd CONTAINING HNO_3 (LOW FEEDING RATE)[7]

ALTERNATIVE ORGANIC REDUCTANTS FOR DENITRATION

K. Gompper
Kernforschungszentrum Karlsruhe GmbH
Institut für Nukleare Entsorgungstechnik

Summary

Besides formic acid and formaldehyde several organic reagents are used to decompose nitric acid. Denitration by use of sugar was first carried out at HANFORD, USA. More than 12 moles nitric acid can be decomposed per mole sucrose. The decomposition is linerarly dependent on the concentration of sugar and the presence of iron increases the reaction rate. On account of the slow reaction, which does not give rise to foaming, a simple uncomplicated design of reaction vessels is possible. Moreover the reagent is very cheap.
At the Karlsruhe Nuclear Research Centre the denitration by use of ethanol was investigated. The nitric acid waste solution can be mixed at room temperature and after heating up to about 80°C the decomposition reaction starts without induction time. In a simulated medium level liquid waste solutiom more than 90% of the initial nitric acid are decomposed within 2 h. The reaction rate depends on the concentration of nitric acid and nitrate salts. The molar ratio of alcohol to nitric acid is between 0.5 and 1. The off gas contains mainly of CO_2 and N_2O. A selective precipitation of transuranium elements during denitration is possible by use of diethyl oxalate. In acid medium this reagent hydrolyzes to give ethanol and oxalic acid, which serve as denitration and precipitation reagents, respectively.

1. Introduction

Nitric acid, one of the most frequently used chemicals in nuclear fuel reprocessing can be decomposed by several methods.

One of the simplest way is the neutralization with hydroxides. This method works very well, but leads to an increased salt concentration in the solution.

The second way consists in acid reduction by dissolution of a metal. The reaction products depend hardly on the initial concentration of the nitric acid. So, gaseous products from hydrogen to nitrous oxides can be formed.

Moreover, the salt concentration increases because the metal is dissolved.

A better way is the elctrochemical decomposition (1). The acid is reduced and NO_2 is generated. But there is a nitric acid limiting concentration below which the nitrate decomposition does not take place and only hydrogen is formed. By addition of copper ions at mg-concentrations this problem can be solved, so that the acid is denitrated down to 0.1 mole/l.

The most interesting methods for denitration are those producing gaseous products without increase of the salt concentration in the solution. The reaction products shall not be explosive or inflammable, and, if possible, not toxic so that they can be released to the environment without further processing.

This goal can be partly reached by the reaction of nitric acid with organic compounds.

Several of these compounds are known which reduce nitric acid to nitrous oxides. During this process the reductants are oxidized to carbon monoxide or dioxide.

The most important organic compounds used in nuclear technology for denitration are

formaldehyde (2) and formic acid (3)

These two reagents will be discussed very comprehensively in other papers to be presented during this seminar.
Moreover, some other reagents were tested or used. These are:

organic complexing agents,

oxalic acid,

glycerin,

sugar (sucrose, fructose, glucose, crude syrup),

ethanol,

diethyl oxalate.

The complexing agents, mainly citric acid, EDTA, oxalic acid and tartaric acid, are effective in the presence of radiation (4).

Glycerin shows a relatively good denitration behaviour, if the nitric acid concentration is high (5). But decreasing the reaction temperature of the acid concentration increases the induction period of the reaction. For example: A 10 M nitric acid at 100°C implies an induction period of 0.5-1 min. But with a 4 M acid the reaction started already after around 8 min, and with a 2 M nitric acid not even before 20 min.

Better results can be achieved with sugar, ethanol and diethyl oxalates. So, in this paper, these reducing reagents will be discussed.

2. Denitration with sugar

Denitratrion experiments with sugar were first carried out at HANFORD, USA by BRAY and co-workers in the early sixties (4). The reasons why they used this reagent were difficulties like e.g., foaming encountered during initial use of formaldehyde in denitrating PUREX waste.

The products of the denitration with sugar are all gaseous and consits of carbon and nitrogen oxides. BRAY proposed the following mechanism starting from sucrose (Fig.1). This shows, that two hydrolysis reactions and at least three oxidation reactions take place before carbon dioxide is formed.

From the summation equations it can be seen, that between 12 and 48 moles of nitric acid may be destroyed by one mole of sucrose consumed (Fig.2).

BRAY tested several parameters which are important to know with a view to the denitration reaction.

Experiments were carried out in which a simulated HAW-solution with approximately 8 moles/l nitric acid was mixed stepwise with increments of a 0.03 molar sucrose solution at 95-105°C. The reaction time for each increment was 4-6 hours.

As a result, it has been shown that the destruction of nitric acid is linearly dependent on the concentration of sucrose added (Fig. 3).

Following HEALY (6), who noted that the presence of ferric or uranyl nitrates catalyse the denitration with formic acid or formaldehyde, the influence of iron ions on the efficiency of the reaction with sugar was tested.

The experiments were performed at 100°C and the molar ratio of nitric acid to sucrose was 24:1, which means that the nitric acid to carbon ratio is 2:1. The concentration of iron, which is added as iron nitrate varied from 0 to 1 mole/l. The reaction time was 24 h.

The higher the concentration of iron had been the more moles of nitric acid per mole of sucrose had been destroyed. Without iron about 12 moles of nitric acid were denitrated, whereas the addition of 1 mole/l iron nitrate increased the efficiency up to about 20 moles nitric acid per mole of sucrose (Fig. 4).

As the reaction time of about 24 h is relatively long, it is of interest to know more about the dependence on time of the reaction rates. The conditions for this experiment were the same as above, only the iron concentration was fixed at 0.8 mole/l.

After 10 minutes about 10-14 moles nitric acid per mole sucrose had already been destroyed. But not before 10 h had it been possible to increase the efficiency to 20 moles (Fig. 5).

Regarding the residual total carbon in solution it was found that after the first 10 minutes only 4% had evolved as oxides. Within 60 min only 40% carbon remained in the solution and after approximately 20 h more than 95% sucrose was destroyed. BRAY pointed out, that all the remaining carbon should be destroyed by radiation in actual waste.

In 1982 Mac DOUGALL, BAYNE and ROBERSON (7) from Oak Ridge National Laboratory published a paper based on the former work of BRAY et. al. They examined nitric acid solutions with up to 16 moles/l. The experiments were carried out both at 100 and 110°C with low iron concentration between 0 and 0.2 moles/l. The nitric acid/carbon ratio was 4:1.

The reaction temperature (100 and 110°C) had no significant influence on the denitration efficiency.

The off-gas composition was detected in an experiment in which the nitric acid/carbon ratio was 2.5:1. At the beginning of the reaction the evolution of NO_2 is violent but decreases steadily whereas the amount of NO increases toward the latter stage of the reaction.

Carbon monoxide was examined too, because large quantities might be an explosion hazard. The CO release remained low, ranging from about 1 to 3 vol%. The amount of CO_2 is low at the beginning of the reaction but increases with time.

The advantage of the denitration by use of sugar is that it does not give rise to foaming because the reaction rate is relatively low. This allows a simple uncomplicated design of reaction vessels. Moreover, the reductant is cheaper compared with formic acid or formaldehyde.

However, because of the slow reaction, the denitration may be the rate determining step in waste treatment.

3. Denitration with ethanol and diethyl oxalate

At the Karlsruhe Nuclear Research Centre, investigations are being performed into the separation of transuranium elements and fission products from medium level liquid waste (8). This includes not only the joint precipitation but also the selective precipitation of TRU-elements and fission products.

Plutonium and americium are precipitated by addition of oxalates and of a carrier. To achieve good crystallisation, crystal growth should be slow. To this end, dimethyl and diethyl oxalates were tested. It is well known that these compounds hydrolyse in an acid medium such as the waste solutions containing nitric acid. So, the oxalate concentration increases slowly. It was found in experiments that the hydrolysis was very slow under the conditions of simulated MLLW at room temperature. To promote this reaction the mixture was heated.

In case of diethyl oxalate, the reaction mixture, which was originally turbid because the ester is not soluble in water, became clear at about 80°C and gas generation started. After cooling down and filtration of the oxalate precipitates the $H^{(+)}$-concentration was determined routinely by titration. Surprisingly, the acid concentration was reduced. This means that a denitration reaction had taken place. Dimethyl oxalate, which was dissolved in ethanol, because it is a solid and not soluble in water, gave the same results with simulated waste solutions.

So it was of interest to investigate the reason of this reaction and to get to know more about the conditions required.

To this end lab-scale experiments were carried out with 200 ml MLLW-simulate and diethyl oxalate, respectively. The two solutions (molar ratio of ester to nitric acid = 1:2) were mixed and heated up to 80°C. When the solution became clear and gas generation started, the temperature was kept between 90 and 100°C. After cooling down, the acid concentration was determined by titration with sodium hydroxide.

The rate of decomposition depended on the reaction time. About 50% of the acid was decomposed in the first 2 hours, but more than 90% within 4-5 hours. Similar results were obtained when the hydrolytic products ethanol and oxalic acid were used instead of the ester.

The oxalic acid however, showed poor denitration efficiency if it was used alone, but the alcohol gave almost the same results as the ester, which suggested that it was the reducing agent.

It was possible to increase the reaction rate by use of a vessel fitted with a reflux condenser. More than 90% of the nitric acid had decomposed within the first 2 hours (Fig. 6).

After the reducing reagent was known, some additional parameters were tested in lab-scale experiments to get more information on this reaction, i.e. on the

- nitric acid concentration,

- sodium nitrate concentration,

- molar ratio of alcohol to nitric acid
  (without sodium nitrate),

- molar ratio of alcohol to nitric acid
  (with sodium nitrate).

The rate of denitration depends on the starting concentration of the acid. A 1 M nitric acid was decomposed by 30%, whereas this was increased up to 40% when the acid was 2 molar and reached 65% at an initial concentration of 3 moles/l and more (Fig. 7).

As to the denitration of salt containing radioactive waste solutions, the influence of sodium nitrate was of interest. The decomposition rate was increased up to a salt concentration of 2.5 moles/l. Higher sodium nitrate contents did not improve the results (Fig. 7).

The ratio of alcohol to nitric acid has also an influence on the denitration rate. Ratios of about 0.5 to 1.5 gave the best results in solutions without sodium nitrate. If nitrate is present even the molar ratio of 0.5 is recommendable (Fig. 8).

So it can be noted that the denitration efficiency depends on the nitric acid and the nitrate concentration. This is comparable with results obtained with other denitration methods based on formaldehyde or formic acid.

The summation equation shows, that 1.2 to 1.5 moles of gaseous compounds were formed per mole of nitric acid (Fig. 9). This was proved by measuring the off-gas volume. It consists mainly of CO_2 and N_2O, but at the end of the reaction also a small amount of NO is formed.

Based on the results already available, the following advantages of this new denitration method can be indicated:

- The off-gas volume is 40-50% lower compared with the formic acid reaction.

- The composition of it shifts slightly to a greater amount of N_2O.

- The required molar ratio of alcohol to nitric acid is 1; formic acid, for example, needs 2:1.

- Alcohol is a very cheap reagent.

- Alcohol and the nitric acid waste can be mixed at room temperature.
  At 75-80°C the reaction starts without an induction time.

- The alcohol, which was not consumed for the denitration can be separated easily from the waste solution by destillation; no additional oxidation reagent like e.g., hydrogen peroxide is required.

Some points have not yet been investigated, namely:

- residual carbon in the solution,

- inflammable compounds in the off-gas,

- influence of heavy and noble metals,

- complete balance sheet of the reaction.

It is evident that they have to be clarified before a final assessment can be made of this new denitration method, especially with regard to its application in nuclear technology.

REFERENCES

/1/ F.Baumgärtner; H.Schmieder
    Use of Elctrochemical Proceses in Aqueous Reprocessing
    of Nuclear Fuels
    Radiochemica Acta 25, 191-210 (1978)
/2/ T.V.Healy
    J. Appl. Chem. 8, 553 (1958)
/3/ S.Drobnik
    German Patent DP 1935273 (1975)
/4/ L.A.Bray; E.C.Martin
    HW-76973 (1962)

/5/ H.Richter; H.Soratin
    Zerstörung des Salpetersäureüberschusses in radioaktiven
    Abfallösungen und Verfestigung des Rückstandes als Alkydharz
    SGAE-2252 (1974)
/6/ T.V.Haley; B.L.Ford
    AERE-C-R-1339 (1954)
    T.V.Healy; B.L.Davies
    AERE-C-R-1739 (1956)
/7/ C.S.Mac Dougall; C.K.Bayne;R.B.Roberson
    Studies on the Reaction of Nitric Acid and Sugar
    Nuclear Technology, 58, 47-52 (1982)
/8/ K.Gompper
    Mittelaktive Abfallösungen
    Abtrennung von Transuranelementen und Spaltprodukten
    atomwirtschaft-atomtechnik 30, 90 (1985)

OXIDYZATION OF SUCROSE
from HW-76973

Fig.1

$12 \ HNO_3 \ + \ C_{12}H_{22}O_{11} \ ==> \ 12 \ CO \ + \ 6 \ N_2O_3 \ + \ 17 \ H_2O$

$12 \ HNO_3 \ + \ 6 \ NO \ ==> \ 18 \ NO_2 \ + \ 6 \ H_2O$

$24 \ HNO_3 \ + \ 12 \ CO \ ==> \ 24 \ NO_2 \ + \ 12 \ CO_2 \ + \ 12 \ H_2O$

$C_{12}H_{22}O_{11} \ + \ 48 \ HNO_3 \ ==> \ 48 \ NO_2 \ + \ 12 \ CO_2 \ + \ 35 \ H_2O$

REACTION OF NITRIC ACID WITH SUGAR

Fig.2   HW–76973

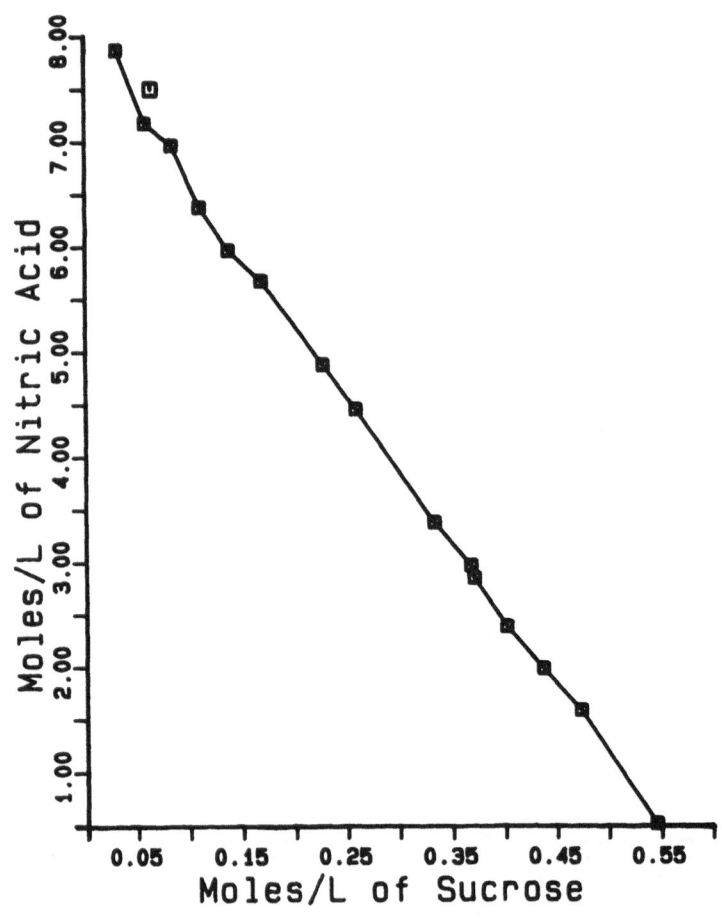

DESTRUCTION OF NITRIC ACID WITH SUCROSE

Fig.3 from HW−7693

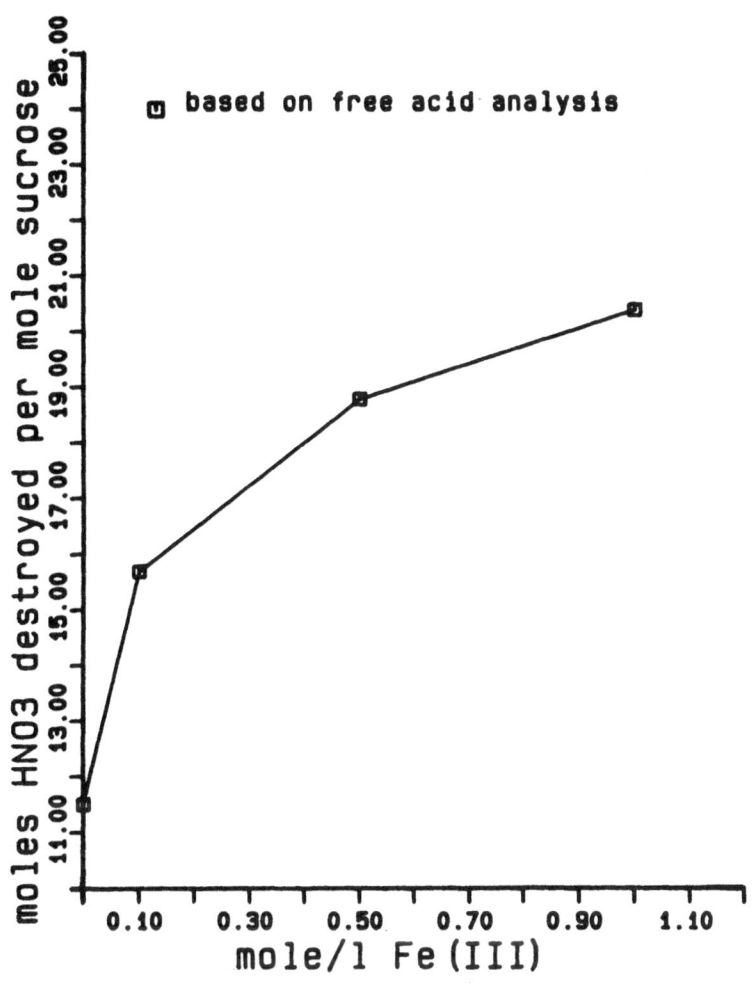

DENITRATION WITH SUGAR AS A FUNCTION
Fig.4 OF IRON CONCENTRATION
from HW–76973

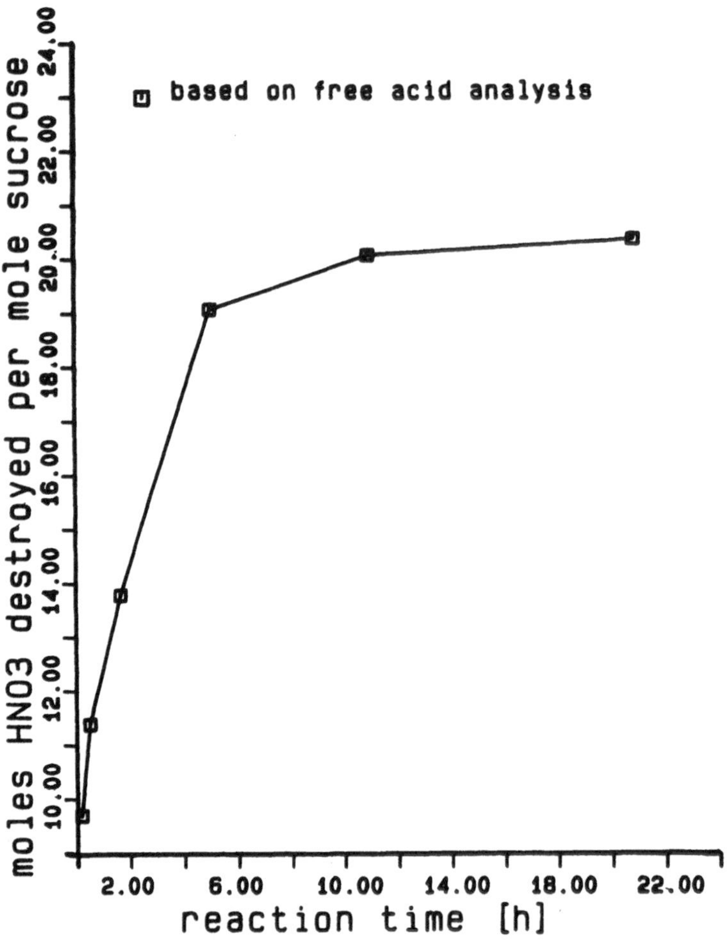

DENITRATION WITH SUGAR AS A FUNCTIO
Fig.5   OF TIME
from HW−76973

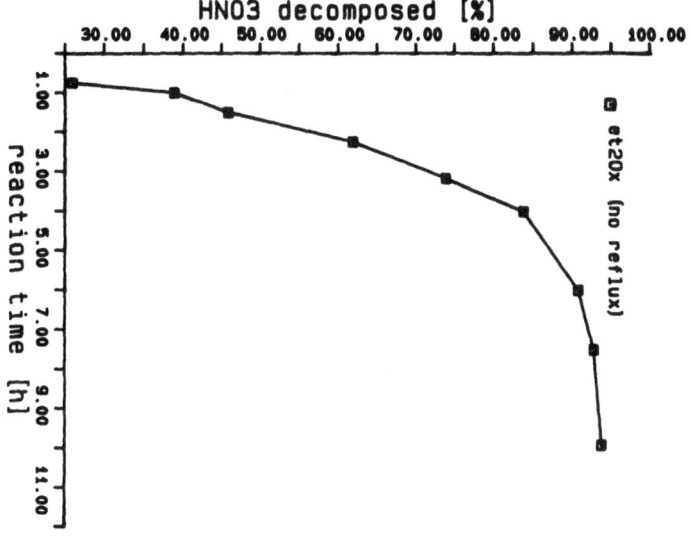

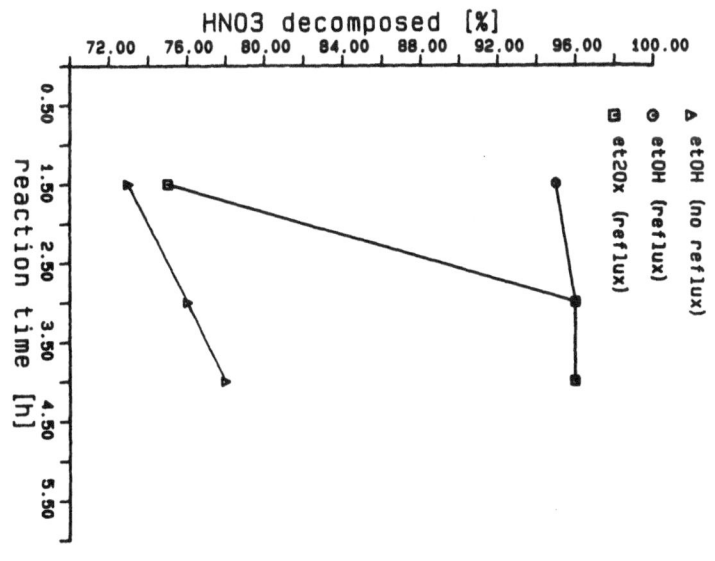

Fig. 6   DENITRATION OF MLLW–SOLUTIONS WITH ETHANOL AND DIETHYLOXALATE

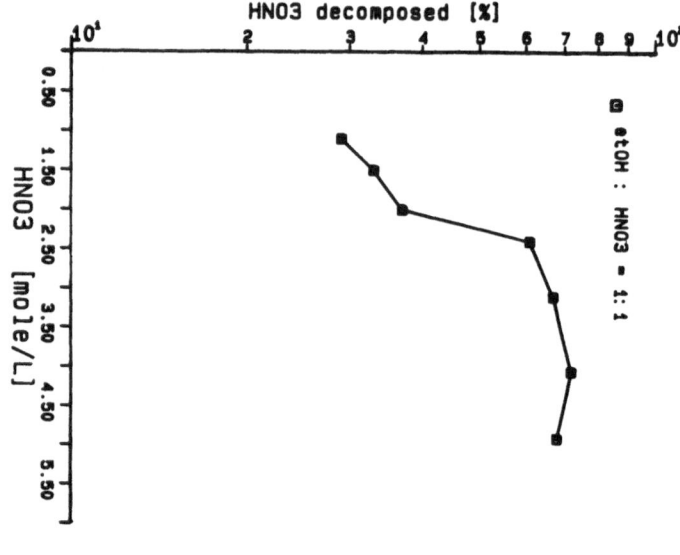

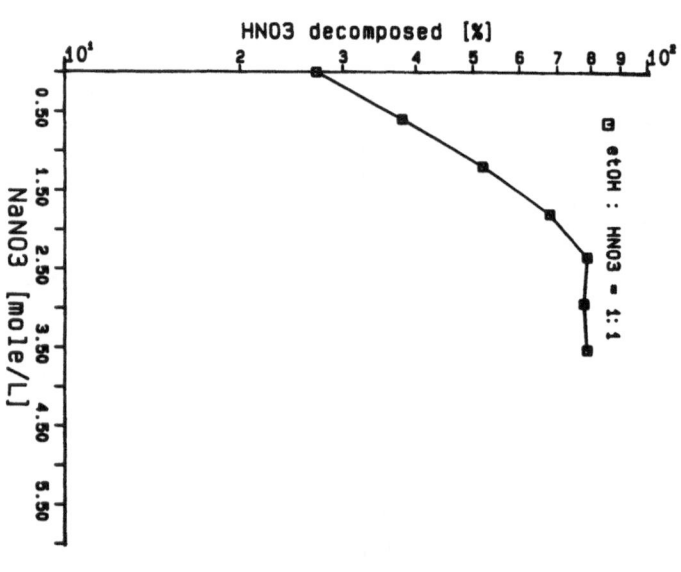

Fig. 7   DENITRATION WITH ETHANOL
INFLUENCE OF THE NITRIC ACID AND SODIUMNITRATE CONCENTRATION

44

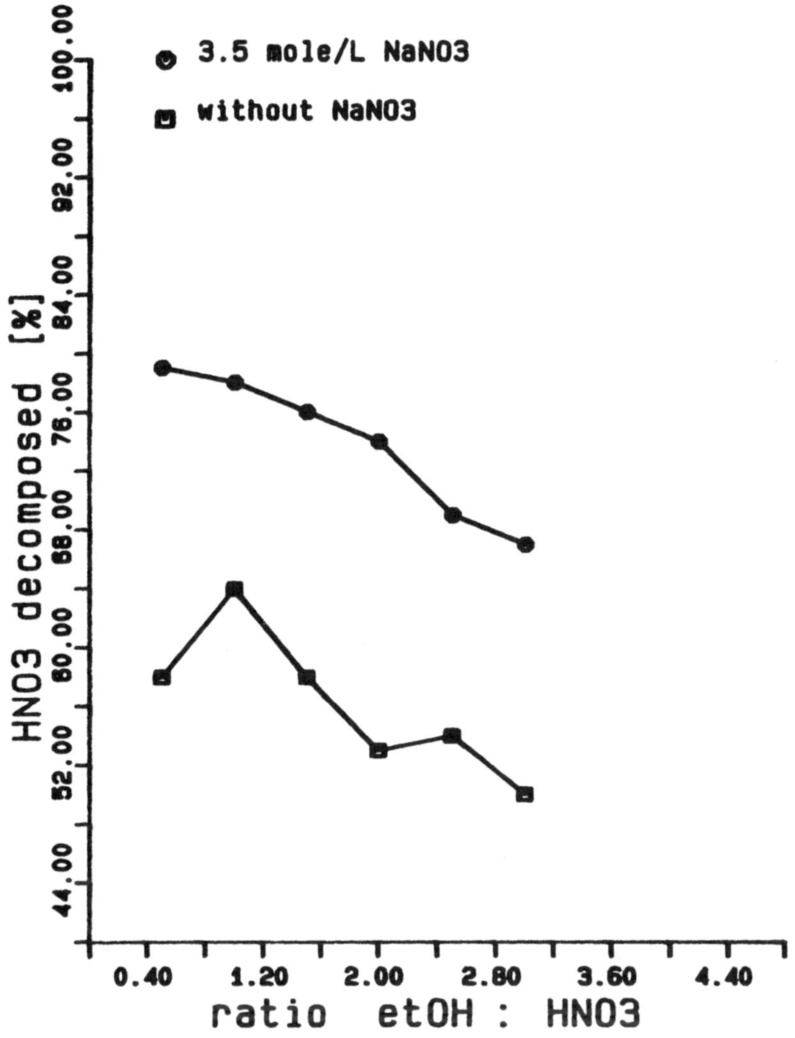

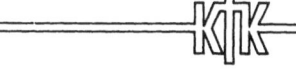

Fig.8   DENITRATION WITH ETHANOL
INFLUENCE OF THE MOLAR RATIO
ETHANOL / NITRIC ACID

$$2\ CH_3-CH_2OH + 6\ HNO_3 \implies 3\ N_2O + 4\ CO_2 + 9\ H_2O$$

$$CH_3-CH_2OH + 4\ HNO_3 \implies 4\ NO + 2\ CO_2 + 5\ H_2O$$

$$CH_3-CH_2OH + 12\ HNO_3 \implies 12\ NO_2 + 2\ CO_2 + 9\ H_2O$$

Fig.9 REACTION OF ETHANOLE WITH NITRIC ACID

ECONOMICAL ASPECTS OF DENITRATION IN
THE MANAGEMENT OF REPROCESSING CONCENTRATE

H. KLONK and M. STEHLE
Deutsche Gesellschaft für Wiederaufarbeitung
von Kernbrennstoffen mbH (DWK), Hannover

Summary

At today's state of the art denitration of HAWC is mandatory prior
to vitrification. Although there is the potential that the ceramic
melter will take HAWC at 5 $\underline{M}$ free nitric acid, the process is not
yet demonstrated. For industrial application corrosion of the off-gas
equipment would be intolerable, and the deliberation of some specific
nuclides at higher acid concentrations will lead to problems with the
off-gas treatment in total.
The DWK-concept for the reprocessing plant at Wackersdorf uses the
HCHO-process. The gaseous reaction products are recombined to nitric
acid and recycled to the high tritium acid recovery system.
Denitration of MAWC is not mandatory but offers the chance of reducing the volume of solidified waste by 340 concrete shielded drums
(400 l) per year. It provides flexibility in the processing of MAW by
not being limited to the maximum salt concentration rather to the
activity limits. For MAWC the HCOOH-process is used.

1. Concept of the reprocessing plant at Wackersdorf (WAW) with respect to denitration

In November 1982 DWK has filed the concept for a 2 t/d reprocessing
plant at Wackersdorf to be approved by the licensing authorities. With
respect to denitration the following process systems of the plant are of
prime interest:
- chop leach head-end
- 1st extraction and partitioning cycle
- 2nd and 3rd Uranium cycle
- 2nd and 3rd Plutonium cycle
- two acid recovery systems
- three solvent clean-up systems
- vitrification of HAWC (ceramic melter)
- cementation of MAW-concentrates
- cementation of LAW-concentrates

Based on an annual throughput of 350 t there will be 280 m³/a of HAWC and
2400 m³/a of MAW.
Denitration processes are foreseen as treatment of the HAWC prior to
vitrification and of MAWC prior to cementation.

2. Criteria for choice of the denitration process

Chemical denitration introduces new chemicals to the process besides
the well known standard chemicals like HNO_3, TBP and kerosine. These chemicals, HCHO and HCOOH, give rise to intense safety considerations and material problems. Looking from the standpoint of operation, denitration must
therefore yield a memorable effect on operating costs, to be considered

for a commercial reprocessing plant. These reductions in operating costs
may be described as
- protection of downstream equipment against enhanced nitric acid
  corrosion
- a positive effect in reducing off-gas activities
- a reduction of the amount of final waste product volume
  by taking full advantage of maximum activity specifications
- a process of withdrawal of excess nitric acid from the plant
  inventory and reducing the amount of HNO_3 to be neutralized
  and unnecessarily treated as radioactive waste

To achieve smooth operation with little risk of upset process conditions
the chosen denitration process must be easy to control and may not result
in foaming nor in forming extense amounts of insoluble matter.

From these criteras DWK chose the HCHO-process for denitration of
HAWC and the HCOOH-process for denitration of MAWC.

As the HCHO-process already is in active operation at AVM at Marcoule
the HCOOH-process still lacks its demonstration of hot operation on a
technical scale. DWK is confident that this process will be fully available
for hot operation of the WAW.

Electrochemical denitration has been considered but ruled out as this
process is believed not to be available for industrial application in the
near future.

3. Denitration of HAWC at WAW

3.1 Management of HAW and HAWC

The raffinate from the 1st extraction cycle (HAW) will be steam-
stripped and concentrated by a factor of 10. The 1W-concentrator is a pot
evaporator operated with continuous feed and batch take off. At the desired fission product concentration water is fed in and excess nitric acid
removed by steam distillation to a concentration of 5 M free HNO_3. Although
vitrification is to be performed right after concentration, 3 buffer
storage tanks (1 as a spare) provide a 3 months decoupling time in case of
up-set conditions in the vitrification. For safe storage of HAWC the HNO_3-
concentration of 5 M is a compromise to prevent enhanced corrosion and formation of precipitates as well.

3.2 Reasons for denitration of HAWC

Prior to vitrification the HAWC will be denitrated to 1.5 M free HNO_3.
At this concentration and regarding the short residence time before feeding
the concentrate to the ceramic melter, formation of precipitates will be
tolerable low.

There are three main reasons to denitrate HAWC:
- Nitric acid vapours at high temperatures are extremely corrosive.
  Although the off-gas equipment will be remotely replaceable it is
  obvious,that frequent exchange of equipment does not comply with
  the requested continuous operation.
- Some fission products, mainly Ruthenium and Technetium tend to form
  volatile comounds to a substantially degree at higher concentrations
  of nitric acid. Although all washing liquids from the off-gas treatment are recycled either to the melter or to the HAW-concentrator,
  it is desirable to keep the amount of these nuclides deliberated
  to the off-gas system small.
- There is a strong evidence, that the calcining reaction of the fission products requires higher temperatures at higher acid concentrations. A low acid concentration will therefore ease the operation of the ceramic melter, resulting in a decreased off-gas

temperature and a smooth vitrification reaction.
As the AVM-process requires the denitration step, the ceramic melter may take HAWC and 5 $\underline{M}$ free nitric acid. This process has not yet been demonstrated. The PAMELA-plant at Mol/Belgium, which is in successful hot operation since October 1985, is designed also for a 1.5 $\underline{M}$ free HNO_3 feed.

3.3 Denitration process and influence on HNO_3 balance

The denitrator is a pot type evaporator heated by pressurized water at 150 °C. One batch of 3,8 m³ of HAWC takes 40 hours for the denitration reaction and 56 hours for the whole process.

With regard to the total HNO_3 balance the denitration process using HCHO does not destroy nitric acid. The gaseous reaction products are recombined to form again nitric acid.

The WAW plant provides two acid recovery systems at a high tritium concentration and a low tritium concentration respectively. Head-end, 1st extraction and HAW concentration and vitrification will be connected to the high tritium cycle only.

As nitrate is continuously removed from the high tritium cycle as Uraneous nitrate and plutonium nitrate solution, a 23 % deficit of the balance has to be replaced by fresh nitric acid for dissolution of fuel elements. Therefore, it is even desirable, that the denitration process for the HAWC returns the nitric acid back to the process. The denitrated and recombined nitric acid is about 60 t of HNO_3 per year, which is 13 % of the balance.

4. Denitration of MAWC

4.1 Management of MAW solutions

The total amount of MAW from the process itself (excluding decontamination during plant shut-down) is 2400 m³/a. Almost 70 % of this volume are slightly acidified washing solutions from the solvent clean-up systems. In addition, depending on the plant availability, frequency of rework operations, quality of process control and unscheduled shut-down periods the amount of MAW can change.

The process MAW, with a calculated acid concentration of 0.16 $\underline{M}$ will be concentrated by a factor of 20. In order to achieve a distillate with as low a nitric acid concentration as possible, a continuously operated concentrator with a fractionating column is used. The capacity is 2 t/h.

The concentrate can be routed either to the denitrator or directly to the cementation adjustment tank. Denitration will be skipped e. g. for decontamination solutions from the remotely operated service and maintenance cell, which are low in salt content and acid concentration.

The concentrates are neutralized with NaOH, adjusted to a salt concentration of 300 g/l at maximum and cementized in 400 l drums provided with a concentrate shielding. This waste product will meet the specifications for low active waste containers suitable for final disposal in the KONRAD repository.

Based on this process two essential specified values for MAW concentrates determine the concentration factor and the total volume of solid waste:
- salt content ($NaNO_3$)   300 g/l
- activity   5.4 E 12 Bq per 400 l drum

4.2 Reasons for denitration of MAWC

The following reasons led DWK in 1982 to provide denitration using the HCOOH process in the processing of MAW concentrates:

- reduction from 980 to 640 drums per year in case of
  denitration of the process MAW
- maximum flexibility to process all kind of aqueous MAW
  solutions, i.e. to take full advantage of the maximum
  permissable activity content
- a low cost possibility to remove excess nitric acid from
  the plant inventory with almost no extra waste volume generated.

4.3 Denitration process and influence on HNO_3 balance

The MAWC denitrator is a pot type evaporator heated by pressurized
water at 135 °C. The required amount of formic acid is brought to boiling
and 950 l of MAWC are fed in during 5 hours. Another 5 hours are required
to finish the reaction, destroy excess of HCOOH with H_2O_2 and concentrate
the resulting solution to the desired salt content. To process the MAWC
arising from continuous plant operation one batch is due every two days.

From the low tritium acid cycle 15 t/a of HNO_3 are lost to the solvent
clean-up waste. Compared to the total loss of 140 t/a - mainly to the
high tritium cycle - and to the production of 515 t/a of recovered acid
these losses are small.

Therefore it is desirable to use a denitration process destroying
nitric acid to gaseous products which can be released via the stack. In
addition, every rework operation, extensive rinsings and decontamination
operation will either increase the plant's inventory of nitric acid or
give waste solutions which are not suitable to be recycled. The excess of
nitric acid can than be economically destroyed rather than neutralized and
cementized.

4.4 Future considerations in the management of MAW

The amount of aqueous MAW generated depends strongly on the maintenance and decontamination techniques. As the high active and medium active
processes at WAW will be maintained remotely, decontamination waste can be
cut down substantially compared to directly maintained plants. Therefore,
the solvent clean-up waste becomes even more dominant.

Any success in reducing MAW from this process i.e. reducing the volume or the sodium content or even avoiding it by other techniques, reduces the primary economical effect for denitration of MAW.

On the other hand, as long as the amount of MAW is not as directly
dependent on the plant throughput of nuclear fuel as it is for HAWC, any
management of liquid wastes - as well as all other wastes - has to be
highly flexible not to affect the availability of the main process. From
the standpoint of operation it seems to be a better idea having the denitration equipment available, maintaining it and operating it occasionally
rather than increasing the solid waste volume by cementizing of essentially
inactive sodium nitrate.

SAFETY ASPECTS ABOUT DENITRATION

T. Schulenberg
Dornier System GmbH
D-7990 Friedrichshafen
Federal Republic of Germany

Summary

The denitration of high and medium active liquid waste with formic
acid is analyzed with respect to the risk of an explosion of the
reaction vessel. The topics which are discussed in this context, are
the formation and reaction of explosive gases, the uncontrolled
reaction of nitric acid with formic acid because of false operation
conditions, the formation and reaction of an organic phase, and the
catalytic effect of traces of rare earths or noble metals. Quantitative results were exemplified. The analysis is based on the reaction
models of Holze et al. [3] and [4].

1. INTRODUCTION

The concept of the nuclear fuel reprocessing plant which was initially planned to be installed near Gorleben (FRG) included a denitration
step for high active liquid wastes (HAW) and medium active liquid wastes
(MAW) after concentration and prior to solidification. It was the aim of
the denitration of HAW to reduce the volatility of Ruthen and the formation of toxic gases during the succeeding vitrification. Moreover, denitration of MAW should reduce significantly the amount of salt within the
solidified waste.

In the German study "Projekt Sicherheitsstudien Entsorgung (PSE)"
[1] the frequency and the consequences of several accidents were analyzed
which were conceivable to occur in a reprocessing plant. In this study
the denitration step was also discussed and the results are summarized in
this paper.

The only accident which might lead to a remarkable release of radioactivity to the public was considered to be an explosion of the reaction
vessel, especially if high active liquid waste is concerned. Therefore,
only those basic events were considered which could finally lead to an
explosion of the vessel.

The denitration method which was chosen in the concept was adopted
from Drobnik [2]. According to this method the nitric acid is reduced
batchwise: formic acid of high concentration (in twofold excess over
nitric acid) is first inserted into the reaction vessel and heated up to
the boiling temperature of 100.6°C. Then the waste solution is gradually
fed into the reactor, and the temperature is controlled by the boiling
point which can increase up to about 107°C depending on the concentrations. The evaporated formic acid is condensed and refed. In a second
phase the reaction mixture is allowed to boil for an additional period
without further addition of liquid waste.

A detailed reaction model for this process was published by Holze et al. [3] in 1979 for MAW, which was extended to HAW by Holze et al. [4] in 1983. The safety analysis applied here is based on these models. However, the basic ideas of this analysis can generally be extended to other reactions.

2. TECHNICAL ASSUMPTIONS

In order to provide some quantitative results some fictitious details of a reaction vessel have been exemplified. The assumed reacor is sketched in Fig. 1. The vessel is upright and cylindrical. It provides suitable room for the denitration of 200 ℓ of nitric acid. The vessel can be heated in the bottom region from outside by a steam condenser of 3 bar and 133°C. Moreover, a vessel design pressure of 3 bar is assumed.

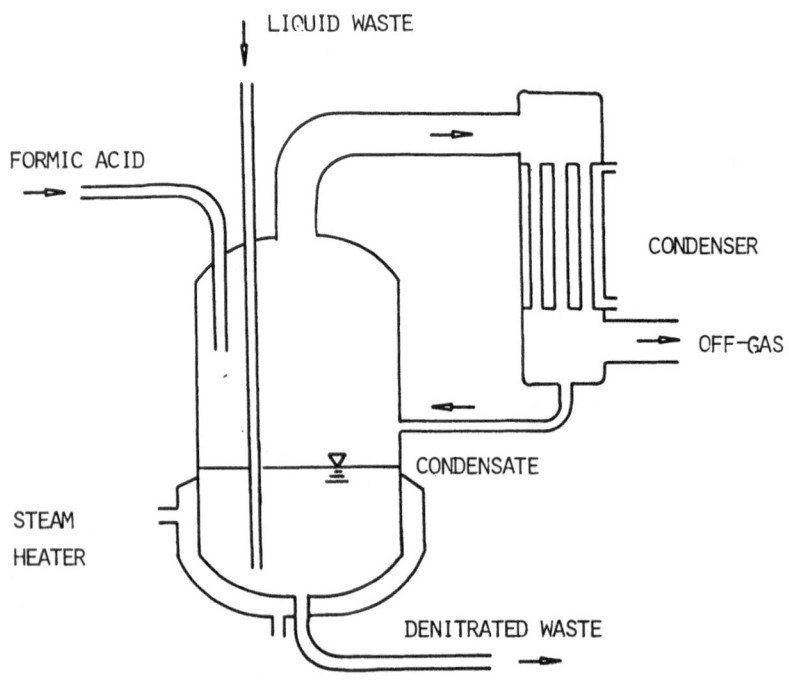

Fig. 1   Sketch of the assumed reactor

In the analysis it has been assumed that the diameter of the off-gas pipe is minimal at the vessel outlet, and that it is reduced there to a cross section of 100 cm^2 due to a vena contracta. This quantity determines the critical flow through the outlet at high gas production rates within the vessel. If the velocity of sound is obtained at this point the mass flux of gas or vapor cannot be increased beyond a critical value G_C. This value is related to the gas properties inside the vessel as

$$G_C = \rho_0 \, a_0 \, (\tfrac{2}{\gamma+1})^{\gamma/\gamma-1}$$

with ρ_0 - density, a_0 - sonic velocity, and γ - polytropic exponent of the gas. From this relation a critical volumetric flow-rate can be derived if the dependence of the gas density and of the sonic velocity on the pressure and the temperature is inserted. Thus, a pressure of 3 bar is predicted if the gas flow-rate should exceed 3000 ℓ/s (normalized to 1 bar and 20°C).

If liquid is also dragged into the off-gas pipe, e.g. if the solution is foaming, the critical mass flux will be reduced significantly. Applying the homogeneous equilibrium model of Wallis [5] a pressure of 3 bar is predicted already if the gas flow-rate exceeds about 1000 ℓ/s (normalized to 1 bar and 20°C) and if the liquid volume fraction in the off-gas pipe is only about 1 %.

3. ESTIMATION OF EXPLOSION LIMITS

A burst of the denitration vessel can occur as a consequence of either of the following three events:

a)   an explosive gas mixture is composed and ignited,

b)   the critical mass flux in the off-gas pipe is exceeded significantly due to a violent, gas producing reaction in the vessel,

c)   a sudden and violent gas production at the bottom of the vessel lifts the whole liquid column; the resulting "water hammer" detonates the upper part of the vessel.

In the following sections these basic events are described in more detail.

3.1 Gas Explosions

The explosion limits of a gas mixture of formic acid, air, and vapor were determined by Kelm et al. [6]. Results are shown in Fig. 2.

In the absence of vapor the gas mixture is explosive if the concentration of formic acid is between 17 and 61 Vol. %. For ignition a minimum temperature of 520°C is required. A maximum explosion pressure of 5.5 bar was calculated in [6], which exceeds the observed explosion pressure of 3.7 bar.

This critical gas composition will always be obtained in the first phase of the denitration when the formic acid is preheated, provided that the empty vessel contained air before. Moreover, a critical gas composition will still be obtained during the first 10 or 20 min of liquid waste addition if about 100 ℓ/h air are inserted for stirring or measurement reasons. However, if the concentration of formic acid in the liquid is reduced by the reaction to less than 20 mol/ℓ, the gas composition becomes subcritical.

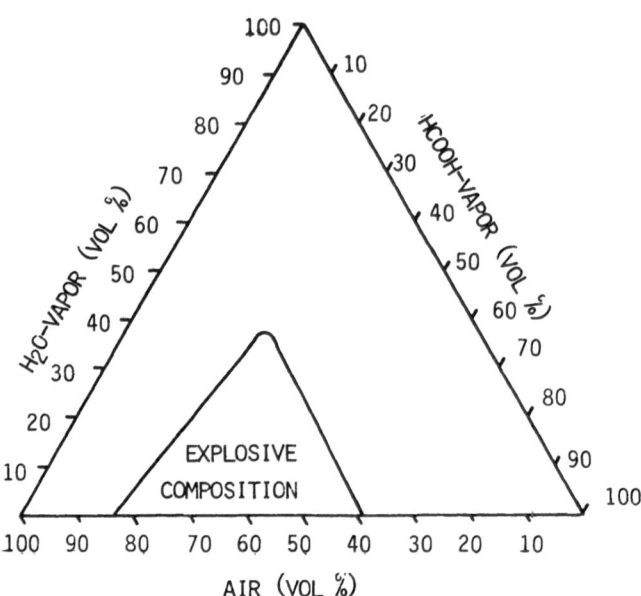

Fig. 2   Explosion limits of a gas containing formic acid; data taken
from [6]

The risk of an explosion is generally low as no ignition sources are
provided, and as the condenser prevents that a critical gas composition
reaches the electrostatical filters of the off-gas system. Moreover, the
succeeding NO_x scrub in the off-gas system is able to trap the escaping
formic acid if the condenser should fail. The risk of an explosion can
further be reduced if nitrogen is used for stirring or measuring, or if
the vessel and the off-gas system are designed for a pressure of about
10 bar or more.

3.2 Violent Denitration

At boiling temperatures the reaction rate and therefore the gas production rate increases linearly with the rate of waste addition. Thus, a
critical mass flux is not expected even if the normal rate of waste addition (i.e. less than 100 ℓ/h) is increased by a factor of ten.

However, if the liquid waste is added already before the boiling
temperature has been obtained, the nitric acid can accumulate because of
an induction period prior to the reaction. This induction period increases with decreasing temperature. It could be approximated with Arrhenius'
law by Holze et al. [3]. Applying these results the accumulated amount of
nitric acid could be calculated. Moreover, it was assumed that the maximum rate of the reaction which follows the induction period is increased

54

by a temperature difference which is about half of the adiabatic temperature increase of the complete denitration reaction [3]. These assumptions
result in a maximum gas production rate after the induction period, which
is plotted as a function of the initial temperature in Fig. 3. The critical mass flux in the off-gas pipe will be exceeded if the initial temperature should be below about 70°C.

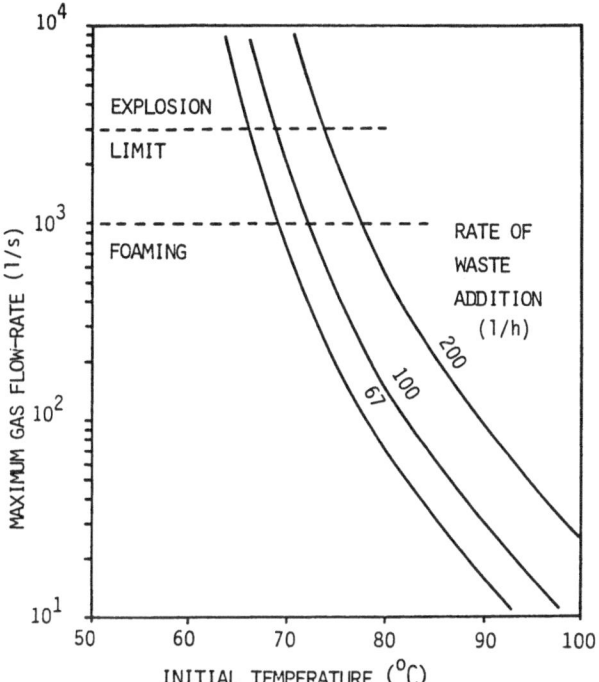

Fig. 3   Maximum gas flow-rate if a liquid waste solution is added to
          non-boiling formic acid

In order to reduce the risk of an explosion it is recommended that
the boiling temperature should be controlled both by temperature measurements and by reflux measurements of the condensate. Moreover, the addition of liquid waste should be slow before the first produced gas has
been detected [3].

3.3 "Red-Oil"

In the concept of the reprocessing plant which is considered here,
it is generally avoided that larger amounts of Tributylphosphate (TBP)
will be transferred to the denitration vessel. However, if the planned
scrubs or steam strippers should fail, an organic phase could probably
separate from the aqueous waste solution which was often designated "Red-
Oil". The exact composition of this organic phase is still unknown. However, complexes of TBP and its degradation products with nitric acid

and heavy metals are considered to play a major role in its composition.

It has been measured by Nichols [7] in 1960 that complexes of TBP with nitric acid decompose violently at temperatures above about 130°C. If this "Red-Oil" were situated at the bottom of the vessel a violent decomposition could rise a "water hammer", such as mentioned above, and could burst the vessel.

In addition to the thermal decomposition of these complexes a chemical decomposition is conceivable which is caused by the denitration of the weakly bound nitric acid. It is an advantage of the denitration method which is analyzed here, that the waste solution is gradually added to the formic acid. By this any "Red-Oil" could decompose under control. However, if the waste solution were first inserted into the vessel and were heated up to the boiling temperature, a larger amount of organic phase could separate in the vessel. If formic acid were added then, the chemical decomposition could rise the temperature of this organic phase and thus could probably initiate a violent thermal decomposition.

However, in order to support this theory further experiments are recommended.

3.4 Oscillatory Denitration

It has been observed by Holze et al. [4] that the denitration of a HAW simulate containing noble metals became unsteady if the rate of liquid addition was too slow or if the liquid addition was interrupted. The reasons for these oscillations which occured then was argued to be a catalytic decomposition of formic acid at elementary Palladium if the concentration of nitric acid became too low. As a consequence of an interruption of waste addition up to 42 Vol. % of Hydrogen were observed in the off-gas [8]. Moreover, large amounts of NH_4^+ were discovered afterwards in the solution, which could probably form ammonium nitrate in a succeeding calcinator.

On the other hand, the reaction rate is hardly controllable during the maxima of oscillations. Although an explosion limit could not be predicted definitely from the few results of Holze et al. [4], it can already be concluded that this denitration method cannot be accepted for HAW if these oscillations cannot be prevented.

4. CONCLUSIONS

Denitration of medium active liquid waste with formic acid as proposed by Drobnik [2] turns out to be a well controllable process under safety aspects. The worst imaginable accident, an explosion of the vessel, can be prevented with a high security factor if the vessel and its instrumentation are designed suitably, because only essential deviations from normal conditions could cause an explosion. Moreover, it is an advantage of this process that no safety problems are expected due to an uncontrolled addition of organic phase. The quantitative safety analysis, however, requires technical details of the vessel which have only been exemplified here.

On the other hand, the application of this process to high active liquid wastes seems to be unsuitable as long as the oscillatory reaction as observed by Holze et al. [4] cannot be prevented.

REFERENCES

[1] HÖRMANN, E. et al. (1985). Projekt Sicherheitsstudien Entsorgung (PSE), Fachband 2, Dornier System GmbH, Friedrichshafen

[2] DROBNIK, S. (1975). Verfahren zum Entfernen von Salpetersäure und/oder Nitrat- und Nitrit-ionen aus wässrigen Abfall-Lösungen, DP 1935273, Gesellschaft für Kernforschung, Karlsruhe

[3] HOLZE, K., FINKE, H.D., KELM, M. and DECKWER, W.D. (1979). Reaction Model for Denitration with Formic Acid of Waste Effluents from Nuclear Fuel Reprocessing Plants, Ger. Chem. Eng. 2, 361-371

[4] HOLZE, K., LI, Z., OSER, B., KELM, M. and DECKWER, W.D. (1983). Auftreten von Oszillationen bei der Denitrierung von salpetersauren schwer- und edelmetallhaltigen Lösungen, Chem. Ing. Techn. 55, 8

[5] WALLIS, G.B. (1980). Critical Two-Phase Flow, Int. J. Multiphase Flow 6, 97-122

[6] KELM, M., OSER, B., DROBNIK, S., BÄHR, W. (1985). Die chemische Denitrierung von salpetersauren schwach- und mittelaktiven Abfall-lösungen mit Ameisensäure, KfK 3948, Kernforschungszentrum Karlsruhe

[7] NICHOLS, G.S. (1960). Decomposition of Tributyl Phosphate Complexes, DP-526

[8] HOLZE, K. (1981). Denitrierung von Abfall-Lösungen aus der Kerntechnik, PhD Thesis, University Hannover

Denitration of reprocessing concentrates by
means of HCHO

S. Halaszovich, S. Dix, R. Harms
Institut für Chemische Technologie
der Nuklearen Entsorgung
Kernforschungsanlage Jülich GmbH

Summary

The denitration of highly radioactive liquid waste and intermediate level liquid waste with formaldehyde have been
investigated. A number of different simulated solutions as
well as real waste from fuel reprocessing have been used
in cold laboratory experiments and in hot experiments
performed with the FIPS II denitrator. Partial or exhaustive reduction of nitric acid can be achieved in a
simple and safe way feeding formaldehyde into the boiling
acidic batch or by simultaneous feeding of acidic waste
and formaldehyde generating mainly NO.

Denitration in a formaldehyde batch with intent to increase N_2O generation is not straight away recommendable
for the following reason. Aqueous formaldehyde usually
contains 10 - 14 % methanol which will react with NO_2
and HNO_2 producing methylnitrite. The concentration of
methylnitrite in the off gas during the first periode of
denitration following the above mentioned procedure might
reach ignition range.

Experiments at ambient temperature supplied the investigations of the reaction mechanism during the induction
periode.

1.   Introduction

Fifteen years ago the Institut für Chemische Technologie der Nuklearen Entsorgung started its research and development activities in the field of solidification of reprocessing concentrates, that time the vitrification of highly radioactive
fission product solutions (HLLW). Like elsewhere in other laboratories also here the decision has been made to take advantage from chemical denitration creating a reducing atmosphere
by this means reducing the discharge of ruthenium which was
known to be volatile as ruthenium tetroxide. Besides the retention of ruthenium another benefit of denitration in the starting step of a solidification process has been expected, namely
the destruction of nitric acid at a low temperature level, thus
reducing the amount of corrosive nitrogen oxides in the hot off
gas line of the melter produced by thermal denitration.

Later, turning to the conditioning of acidic intermediate level
liquid wastes (ILLW), the third advantage devoted to denitration
became obvious. The involved procedures often require elevated

pH values. Usually the pH will be adjusted by addition of sodium hydroxide consequently increasing the salt contant in the waste solution. Using a suitable reducing agent, nitric acid will be decomposed to gaseous products and less salt will be produced by the addition of sodium hydroxide.

There are several substances usable for denitration like sugar, formic acid and formaldehyde. Each of them has specific advantages and disadvantages.

In Jülich aqueous formaldehyde solution has been chosen as promising reducing agent. It has a higher reducing power as for example formic acid, no solids will be produced by its reaction with nitric acid, being a liquid it can easily be transfered and metered and it is non-corrosive. These advantages were expected to compensate the fact that the waste volume will be enlarged by the small portion of water added with the aqueous formaldehyde solution.

2.   Experimental equipment

The small scale experiments have been performed in a laboratory equipment using a three-necked flask with one litre batch.

For large scale and hot experiments the denitrator of the FIPS II equipment has been used /1/. Figure 1 shows the denitration step installed in the hot cell. Figure 2 shows the simplified flow sheet. The vessel is a pot type evaporator with a Raschig ring column and a reflux condenser mounted on the top of it. The reflux condenser can be operated alternatively if no further concentration of the batch volume is desirable. For simultaneous concentration and denitration the second condenser in the off gas line is taken into operation. A condensate collector and two NO_2 scrubbers also belong to this unit. All vessels and tubes are made of stainless steel, the denitrator itself is made of INCOLOY 825.

The denitration unit is equiped with detecting elements for level, density, temperature, pressure and off gas flow rate. A cooled by-pass leads off gas through infrared analyzers to measure the concentration of CO_2, NO_2, NO and N_2O in the off gas. All data are recorded. It is important to recognize that the off gas passes the Raschig ring column and the condenser before it enters the analyzers. Due to absorption the composition measured is not the same as generated by the reaction in the denitrator. Specially the NO_2 and NO values are affected according to the equation

$$3\ NO_2 + H_2O \rightleftarrows 2\ HNO_3 + NO$$

That means the NO_2 amount produced is higher and the amount of NO lower than calculated on the basis of indicated figures. For this reason it is not possible to draw conclusions accurately on the portions of the different gaseous products as generated by the reaction. However this is not only a disadvantage of the FIPS II equipment but a general drawback to be considered where-ever gaseous samples necessarily change their composition on their way from the place they are generated to the analyzer.

3.   Experimental programme

The experimental programme has been directed to the following aims:

-   Investigation and comparison of three alternative ways
    of procedures for denitration:
    • simultaneous feeding of HLLW and formaldehyde
      into the denitrator,
    • feeding of formaldehyde into the HLLW or ILLW
      batch,
    • feeding ILLW into the formaldheyde batch.
-   Investigation of the influence of methanol on denitration.
-   Investigation of the influence of the waste composition
    and of the different ways of procedure on the off gas
    composition.
-   Determination of induction times.
-   Observation of autocatalytic events during the induction
    period .

4.   Composition of the waste solutions

The tables 1 to 6 show the composition of the simulated ILLW and HLLW solutions for cold experiments and the composition of the real ILLW used in hot experiments. The real ILLW was produced in the reprocessing plant WAK in Karlsruhe and pre-concentrated in the HDB, the department for decontamination and waste treatment in the Karlsruhe Nuclear Research Centre.

As it will be shown later, there is a significant difference between the composition of the gas generated during the denitration of a simple simulate like MAW-S/07 and those generated during the denitration of more realistic ones like MAW-S/08 or during denitration of genuine ILLW.

5.   Denitration of ILLW

5.1 Experimental conditions

The experimental conditions using the FIPS II denitrator were as follows:

-   The total volume of the denitrator is 70 l.
-   The maximum batch volume is 10 l.
-   Using the top condenser 90 % reflux can be adjusted.
-   The pressure is adjusted to 960 mbar.
-   Feeding is started at boiling temperature or at 80°C.

60

- The temperature of the batch during denitration is 100°C.
- The prefered feed rate for ILLW is 10 l/h
  (10 - 14 mol HNO_3/h).
- Two different eqùous formaldehyde solutions are used:
  • 13 molar with 10 - 14 % methanol,
  • 10 molar with less than 1 % methanol.
- The prefered feed rate for formaldehyde is 1 l/h.
- The normalized throughput for nitric acid under these
  conditions is 1 - 1,5 mol HNO_3/h·l batch.
- After the feeding period the temperature is usually
  held at boiling temperature for one hour.

5.2 Experimental results

Under the above described conditions reaction started 3 to 5
minutes after having started the feed at boiling temperature
but already after 1 - 1,5 minutes when feed had been started
at 80°C.

The pH value of denitrated simulates is 3 - 3,5 whereas pH 2 was
measured in denitrated real ILLW WAK-MAW-Concentrate-01 and
pH 2,6 in case of WAK-MAW-Concentrate-02.

No residual nitric acid could be found by titration in simulates
after denitration. The titration of residual nitric acid in real
ILLW was not possible due to handling difficulties caused by the
radiation.

The specific formaldehyde consumption mol HCHO/mol HNO_3 was
found to be 0,7 - 0,8 if formaldehyde was fed into the nitric
acid solution and 0,9 - 1 if the nitric acid solution was fed
into formaldehyde.

In contrary to early publications /2/ methanol turned out to
have an effect on denitration. A discrepancy between the total
flow rate indicated by the flow meter and the flow rate calculated by addition of the measured components (fig. 3) shows
that there must be a further gas component besides CO_2, NO_2,
NO and N_2O. As figure 3 shows, this effect can in particular
be observed during the first 20 minutes of denitration if the
acidic solution is fed into formaldehyde. Until that time only
formaldehyde stabilized with 10 - 14 % methanol as commercially
available had been used. Laboratory tests showed that during
this periode of 20 minutes the off gas leaving the off gas line
is inflammable.

As methanol and nitric acid are mixed in the batch, the
formation of an ester by reaction of the acid with alcohol could
be expected. Literature /3/ shows that a reaction between
methanol and nitrogen dioxide and dinitrogen tetroxide forms
methylnitrite (CH_3ONO). The formation of it by reaction of
methanol with nitrous acid is also described /4/. Nitrous acid
is present during denitration as an intermediate product /5/.

The presence of methylnitrite in the off gas could be demonstrated by UV spectrometry. It is not only produced if the
acidic waste solution is fed into the formaldehyde batch but
also if the procedure is performed inversely. However the
concentration of it remains below ignition range.

In figure 4 there is a comparison between the concentration of
methylnitrite during both procedures. The HCHO-batch graph in
this figure and the corresponding graphs in figure 3 indicate
that the formation of methylnitrite obviously governs the
reaction, because no other products are to be observed while
methanol is converted.

Since the methylnitrite production consumes nitrogen dioxide the
amount of this gas component generated during denitration must be
higher than the amount calculated with the measured concentration.

Figure 5 shows the balance of the conversion of nitric acid.
Though the figures lack accuracy to a certain extent due to the
uncertainties in gas analysis mentioned before they clearly show
the influence of procedure and that of the composition of the
waste solution.

If a simple simulate without fission products and corrosion
products is denitrated a larger amount of HNO_3 will be converted
to N_2O specially if denitration is performed in the formaldehyde
batch. This is also true using pure nitric acid. It is obvious
that also during denitration with formaldehyde considerable
amounts of N_2O will be produced via formic acid which is an
intermediate product formed by the first step of oxidation of
formaldehyde.

The presence of fission products will drasticaly shift the
balance towards NO production. This effect is less pronounced if
formaldehyde is fed into the acidic solution. In that case
conversion mainly goes to NO denitrating both types of solutions.
This is also shown in figure 6 comparing the amounts of nitric
oxides as they are present in the off gas. Here is N_2O even less
important compared with the previous figure, where the fact had
to be taken into account, that two moles of HNO_3 are destroyed
to form one mol N_2O.

The results obtained with real MAW underline the results of the
cold experiments. As figure 7 shows the total amount of NO produced during the denitration of WAK-MAW-Concentrate/01 is five
times higher than the amount of N_2O but during denitration of
WAK-MAW-Concentrate/02 which has a higher fission product and
corrosion product content (see table 3) the amount of NO is ten
times higher than that of N_2O. This tendency is in good
accordance with that of the cold experiments.

6.   Denitration of HLLW

6.1 Experimental conditions

The experimental conditions using the FIPS II denitrator were as
follows:

- The total volume of the denitrator is 70 l.
- The maximum batch volume is 10 l.
- The pressure is adjusted to 960 mbar.
- Feeding is started at boiling temperature.
- The batch temperature during denitration is 100°C.
- Two different equous formaldehyde solutions are used
  - 13 molar with 10 - 14 % methanol,
  - 10 molar with less than 1 % methanol.
- The feed rate for simultaneous feeding is
  - 2,5 l HAWC/h and 1 l HCHO/h,
  - 5   l SBR/h and 2 l HCHO/h.
- The feed rate for formaldehyde into the LEWC batch
  is 1 l/h.
- The normalized throughput is
  - for HAWC: 2   mol HNO_3/h·l batch,
  - for SBR: 4   mol HNO_3/h·l batch,
  - for LEWC: 1,5 mol HNO_3/h·l batch.
- The post reaction time while the batch is held at
  boiling temperature is 1 - 2 h.

6.2 Experimental results

Figure 8 shows the balance of nitric acid conversion during
denitration of HLLW compared with the denitration of 5 molar
nitric acid and with ILLW simulate MAW-S/07. The conversion to
NO is supervailing, no matter if the salt content in HLLW is low
(HAWC) or high (LEWC). The ratio between the different nitric
oxides is nearly the same, there is no significant difference
between simultaneous feeding (HEWC) and formaldehyde feeding
(LEWC). This is also true for the denitration of 5 molar pure
nitric acid (no salts) and MAW-S/07 (high salt content). In all
cases the same general formula governs the reaction, influenced
by side reactions shifting the balance to N_2O in case of
solutions without fission products.

The figure also shows another effect, which is probably also
there in case of the denitration of ILLW but it is not as
evident as in case of HLLW. Obviously not only the free nitric
acid has been destroyed but also a portion of the nitrates has
been converted into nitric oxides. Therefore the addition of the
single HNO_3 portions which were calculated from the amounts of
the different nitric oxide species produced during denitration
exceeds 100 % as the initial amount of HNO_3 has been taken
as 100 % basis. This is specially evident in case of LEWC which
contains comparatively large amounts of iron nitrate and aluminium nitrate.

63

Figure 9 shows the generation rate of CO_2 and the nitric oxide species during the denitration of HLLW (SBR). The process is running smoothly and due to simultaneous feeding of HLLW and HCHO and due to the fact that the batch volume is kept constant, also almost stationary. The conditions will become stable sooner if the batch volume is small compared with the total amount to be put through because constant acid and formaldehyde concentrations will sooner arise.

As figure 9 shows the tendency of NO_2 formation differs from that of the formation of N_2O. More NO_2 than N_2O will be produced first but this will change as HNO_3 concentration will decrease gradually. The influence of two different side reaction is obvious. In the first half it could be

$$2 \ HNO_3 + HCHO \rightarrow 2 \ NO_2 + HCOOH + H_2O$$

and in the second half

$$2 \ HNO_3 + 2 \ HCHO \rightarrow N_2O + 2 \ CO_2 + 3 \ H_2O.$$

The main reaction could be

$$4 \ HNO_3 + 3 \ HCHO \rightarrow 4 \ NO + 3 \ CO_2 + 5 \ H_2O$$

because NO is the prevailing nitric oxide species during the whole time.

In contrary to the denitration of ILLW no induction time could be measured during the experiments with HLLW. Reaction starts within seconds in the HLLW batch with an acid concentration of 4 - 5 mol/l.

In contrary to denitration in a batch of the reducing agent /6/ oscillations in the reacting mixture never could be observed during the experiments.

7.   Autocatalytic events during the induction period

Autocatalytic processes and the formation of intermediate species during the induction time can be observed in mixtures of nitric acid and reducing agents because the reaction rate at ambient temperature is low enough. This has been done for the mixture of 14,5 molar HNO_3 and concentrated HCOOH by Cecille /7/.

Very similar effects can also be seen using HCHO but there is a significant difference with respect to the reactivity of the mixture. While reaction starts after few minutes in a mixture of 1 ml nitric acid and 5 ml formic acid, there is no reaction even after one day in a mixture of 14,5 molar nitric acid and 13 molar formaldehyde mixed in the same proportion. However if a small portion of HCHO is given into an excess of 14,5 molar nitric acid reaction will start within seconds and few minutes as it is shown in figure 12.

Figures 10 and 11 show the different stages as the reaction starts and proceeds in a mixture of 5 ml 14,5 molar nitric acid and 1 ml of 13 molar formaldehyde. In one case nitric acid and formaldehyde had been mixed in the other a formaldehyde layer had been added.

The formation of a transparent yellow substance can be observed on the surface of the mixture or at the phase boundary. The colour turns to green and the reaction starts at the same time producing NO_2. During the following reaction a green disc moves down releasing greenish trails to the surface where NO_2 is set free. In case of the mixture the disc moves to the bottom. The liquid remains yellowish because there is a lack of formaldehyde and the conversion of nitric acid will not be completed. In case of the formaldehyde layer the green disc is more distinct. As the reaction stops the disc also stops moving down and there is a yellow one under it. The remaining liquid at the bottom is colourless the liquid above the discs greenish.

8.   Conclusions

Two different procedures can be recommended for the denitration with formaldehyde, first the simultaneous feeding of the waste solution and of formaldehyde keeping the batch volume constant by evaporation, if the waste solution has a low salt content, usually HLLW, and second the denitration with reflux condenser feeding the formaldehyde into the batch of waste solution, if the waste has a high salt content, usually ILLW.

Except pure nitric acid, there is no significant difference between these two procedures concerning the off gas composition. In both cases NO is the predominating nitric oxide species.

The ratio NO/N_2O depends on the contents of fission products and corrosion products in the waste solution. Increasing contant of these species will shift the balance towards NO.

Solutions with a low salt content can further be concentrated keeping the batch volume constant by evaporation while feeding both components waste and formaldehyde simultaneously into the batch.

The induction time depends on the local concentration of an intermediate product, probably HNO_2. The accumulation of this product will be enhanced by increasing. HNO_3 concentration and by increasing temperature but slowed down by stirring for example due to boiling.

Concerning safety the simultaneous but separate feeding of both co-reactants has an advantage over other procedures because overdosage of one of the reactants does not result in violent reaction. The overdosage of both streams is much more improbable.

Solutions which can not be concentrated further should be denitrated by feeding formaldehyde into the acidic batch to keep the concentration of methylnitrite below ignition range. During denitration in the formaldehyde batch even with reduced methanol content methylnitrite will preferably be produced first until methanol is consumed.

Formaldehyde has a high reducing power as it is first oxidized to formic acid and then the formic acid to carbon dioxide. This is an advantage because first the specific consumption $HCHO/HNO_3$ is low and second less carbon dioxide is produced whereby the off gas system is relieved to some extent.

Denitration ends up at low acidity in a comparatively short time. There is no nitric acid and only negligible amounts of formic acid or formaldehyde left if procedures with formaldehyde feed are used. The feed can be stoped as gas evolution and pressure indicate the end of reaction. Therefore contrary to denitration in the batch of reducing agent it is possible to add the proper quantity of formaldehyde even if the nitric acid content of the waste solution is not well known.

The pH values found after denitration range between 2 and 3,5. The value depends on the concentration of fission products and corrosion products in the waste solution. Increasing concentration results in higher pH value.

Experiments at ambient temperatures show that mixtures with an excess of nitric acid are more reactive than mixtures with an excess of formaldehyde. Concerning safety this is an advantage of procedures feeding formaldehyde into the acidic batch.

Many parameters like waste composition, changing concentrations, design characteristics of the reaction vessel have an influence on the type of reaction in a large scale process. Therefore results from laboratory experiments can not be taken directly for the design of larger plants as for example a stoichiometric equation describing the reaction in a narrow range of concentration taken for the calculation of the off gas composition.

REFERENCES

/1/ Dix, S.,
FIPS II-Denitrierungsanlage, Aufbau und Betriebserfahrungen,
Jül-Spez-324, 1985

/2/ Morris, J.B.,
The reaction of nitric acid with formaldehyde,
AERE-CE/R-1490, Harwell (1954)

/3/ Yoffe, A.D., Gray, P.,
Esterification by Dinitrogen Tetroxide, Journal of the
Chemical Society, London (1951), S. 1412

/4/ Beilstein, 4. Auflage, 1. Ergänzungswerk, Band 1,
S. 141

/5/ Longstaff, J.V.L., Singer, K.,
The Kinetics of Oxidation by Nitrous Acid and Nitric Acid,
Part II, Oxidation by Formic Acid in Aqueous Nitric Acid,
Journal Chem. Soc. 1954, S. 2610

/6/ Holze, K.,
Denitrierung von Abfallösungen aus der Kerntechnik,
Dissertation, Universität Hannover (1980)

/7/ Cecille, L.,
Chemical reactions involved in the denitration process
with HCOOH and HCHO
(This seminar)

COMPOSITION OF MAW-S/07
========================

Comp.	Conc. (g/dm3)	used Compound	Conc. (g/dm3)	Oxide	Conc. (g/dm3)	Part %
NaNO3	300	NaNO3	300	Na2O	109,381	100
Al	--	Al(NO3)3 x 9 H2O	--	Al2O3	--	--
Ca	--	Ca(NO3)2 x 4 H2O	--	CaO	--	--
Cr	--	Cr(NO3)3 x 9 H2O	--	Cr2O3	--	--
Cs	--	CsNO3	--	Cs2O	--	--
Cu	--	Cu(NO3)2 x 3 H2O	--	CuO	--	--
Fe	--	Fe(NO3)3 x 9 H2O	--	Fe2O3	--	--
K	--	KNO3	--	K2O	--	--
Mg	--	Mg(NO3)2 x 6 H2O	--	MgO	--	--
Mn	--	Mn(NO3)2 x 4 H2O	--	MnO	--	--
Mo	--	Na2MoO4 x 2 H2O	--	MoO3	--	--
Na	--			Na2O	--	--
Ni	--	Ni(NO3)2 x 6 H2O	--	NiO	--	--
Ru	--	Ru(NO)(NO3)3 -Lsg. 10 % Ru	--	RuO2	--	--
Sr	--	Sr(NO3)2	--	SrO	--	--
Zn	--	Zn(NO3)2 x 4 H2O	--	ZnO	--	--
HNO3	1 mol/l					
				Total	109,381	100

Table 1:   Composition of the synthetic ILLW MAW-S/07

68

COMPOSITION OF MAW-S/08

Comp.	Conc. (g/dm3)	used Compound	Conc. (g/dm3)	Oxide	Conc. (g/dm3)	Part %
NaNO3	300	NaNO3	300	Na2O	109,381	94,06
Al	0,23	Al(NO3)3 x 9 H2O	3,198	Al2O3	0,435	0,37
Ca	1,5	Ca(NO3)2 x 4 H2O	8,838	CaO	2,099	1,81
Cr	0,08	Cr(NO3)3 x 9 H2O	0,616	Cr2O3	0,117	0,1
Cs	0,3	CsNO3	0,44	Cs2O	0,318	0,27
Cu	0,15	Cu(NO3)2 x 3 H2O	0,57	CuO	0,188	0,16
Fe	0,38	Fe(NO3)3 x 9 H2O	2,749	Fe2O3	0,543	0,47
K	0,08	KNO3	0,207	K2O	0,096	0,08
Mg	0,75	Mg(NO3)2 x 6 H2O	7,912	MgO	1,244	1,07
Mn	0,08	Mn(NO3)2 x 4 H2O	0,366	MnO	0,103	0,09
Mo	0,38	Na2MoO4 x 2 H2O	0,958	MoO3	0,57	0,49
Na	0,18			Na2O	0,245	0,21
Ni	0,08	Ni(NO3)2 x 6 H2O	0,396	NiO	0,102	0,09
Ru	0,15	Ru(NO)(NO3)3 -Lsg. 10 % Ru	1,5	RuO2	0,197	0,17
Sr	0,3	Sr(NO3)2	0,725	SrO	0,355	0,31
Zn	0,15	Zn(NO3)2 x 4 H2O	0,6	ZnO	0,187	0,16
Zr	0,08	ZrO(NO3)2 x X H2O (ca. 45 % ZrO2)	0,24	ZrO2	0,108	0,09
HNO3	1 mol/l					
				Total	116,288	100,00

Table 2:   Composition of the synthetic ILLW MAW-S/08

```
##########################################################################################
#                                              #                                         #
#          WAK-MAW-CONCENTRAT/01               #          WAK-MAW-CONCENTRAT/02           #
#                                              #                                         #
##########################################################################################
#                                              #                                         #
#     Content of cations : 3.7  mol/dm3        #     Content of cations : 4.21 mol/dm3    #
#     Density            : 1.19  kg/dm3        #     Density            : 1.19  kg/dm3    #
#     Hydrogen-ion conc. : 1.17 mol/dm3        #     Hydrogen-ion conc. : 3.03 mol/dm3    #
#     pH                 : 1.0                  #     pH                 :  0              #
#                                              #                                         #
##########################################################################################
#                                              #                                         #
#        Main γ-activity distribution          #        Main γ-activity distribution      #
#        ==============================        #        ==============================    #
#                                              #                                         #
#        Co  60     0.05  GBq/dm3              #        Co  60     0.11  GBq/dm3           #
#        Zr  95      -    GBq/dm3              #        Zr  95     0.11  GBq/dm3           #
#        Rh 106     1.66  GBq/dm3              #        Rh 106     2.45  GBq/dm3           #
#        Sb 125     0.74  GBq/dm3              #        Sb 125     0.76  GBq/dm3           #
#        Cs 137     4.17  GBq/dm3              #        Cs 137    56.60  GBq/dm3           #
#        Ce 144     1.61  GBq/dm3              #        Ce 144     0.55  GBq/dm3           #
#                                              #                                         #
##########################################################################################
```

Table 3:   Excerpt of analytical data for genuine ILLW

COMPOSITION OF DWK-HAWC-S/02 (15dm3-Mixture dissolved in 27dm3)
===

Comp.	Conc. (g/dm3)	Mass (g)	used Compound	Mass (g)	Oxide	Mass (g)	Conc. (g/dm3)
Rb	0,488	7,32	RbNO3	12,63	Rb2O	8	0,533
Sr	1,155	17,33	Sr(NO3)2	41,85	SrO	20,49	1,366
Y	0,659	9,89	Y(NO3)3 x 6 H2O	42,59	Y2O3	12,55	0,837
Zr	5,461	81,92	ZrO(NO3)2 x x H2O (45%ZrO2)	245,89	ZrO2	110,65	7,377
Ru	3,37	50,55	Ru(NO)(NO3)3 -sol. 10 % Ru-content!	505,5	RuO2	66,55	4,437
Rh	0,574	8,61	Rh(NO3)3 -solution 10 % Rh-content!	86,1	Rh2O3	10,62	0,708
Cd	0,158	2,37	Cd(NO3)2 x 4 H2O	6,5	CdO	2,71	0,181
Cs	3,542	53,13	CsNO3	77,92	Cs2O	56,33	3,755
Ba	2,654	39,81	Ba(NO3)2	75,76	BaO	44,45	2,963
La	1,9	28,5	La(NO3)3 x 6 H2O	88,84	La2O3	33,42	2,228
Ce	3,704	55,56	Ce(NO3)3 x 6 H2O	172,18	CeO2	68,25	4,55
U	2,597	38,96	>				
Np	0,72	10,8	>				
Pu	0,037	0,56	>Ce(NO3)3 x 6 H2O	122,76	CeO2	48,66	3,244
Am	1,002	15,03	>				
Cm	0,153	2,3	>				
Pr	1,776	26,64	Pr(NO3)3 x 5 H2O	78,84	Pr6O11	32,19	2,146
Nd	6,158	92,37	Nd(NO3)3 x 5 H2O	269,18	Nd2O3	107,74	7,183
Pm	0,024	0,36	Nd(NO3)3 x 5 H2O	1,03	Nd2O3	0,41	0,027
Sm	1,346	20,19	Sm(NO3)3 x 5 H2O	57,25	Sm2O3	23,41	1,561
Eu	0,279	4,19	Eu2O3	4,85	Eu2O3	4,85	0,323
Gd	0,246	3,69	Gd(NO3)3 x 5 H2O	10,17	Gd2O3	4,25	0,283
Si	---	---	Sodium silicate	----	SiO2	---	--
Na	---	---	Sodium silicate	----	Na2O	---	--
Cr	0,562	8,43	Cr(NO3)3 x 9 H2O	64,88	Cr2O3	12,32	0,821
Mn	0,063	0,95	Mn(NO3)2 x 4 H2O	4,34	MnO	1,23	0,082
Tc	1,26	18,9	Mn(NO3)2 x 4 H2O	47,92	MnO	13,54	0,903
Fe	2,158	32,37	Fe(NO3)3 x 9 H2O	234,17	Fe2O3	46,28	3,085
Ni	0,312	4,68	Ni(NO3)2 x 6 H2O	23,18	NiO	5,95	0,397
PO4	0,382	5,73	H3PO4 (85%ig)	6,96	P2O5	3,32	0,221
Cl	0,095	1,43	HCl-Titrisol 0,1n	402 ml			
F	---	---	HF (40%ig)	----			
Hf	---	---	HfO2	----	HfO2	---	--
Al	0,124	1,86	Al(NO3)3 x 9 H2O	25,86	Al2O3	3,51	0,234
Mg	0,124	1,86	Mg(NO3)2 x 6 H2O	19,62	MgO	3,08	0,205
Ca	0,124	1,86	Ca(NO3)2 x 4 H2O	10,96	CaO	2,6	0,173
Se	0,076	1,14	Se	1,14	SeO2	1,6	0,107
Pd	2,463	36,95	Pd(NO3)2 -solution 10 % Pd-content!	369,45	PdO	42,5	2,833
Ag	0,1	1,5	AgNO3	2,36	Ag2O	1,61	0,107
K	0,124	1,86	KNO3	4,81	K2O	2,24	0,149
Na	0,106	1,59	NaNO3	5,88	Na2O	2,14	0,143
Te	0,895	13,43	Te	13,43	TeO2	16,8	1,12
Sn	0,085	1,28	Sn	1,28	SnO2	1,63	0,109
Sb	0,019	0,29	Sb	0,29	Sb2O3	0,35	0,023
Mo	5,824	87,36	Mo	87,36	MoO3	131,07	8,738
HNO3	5 mol/l						
					Total	947,3	63,152

Table 4:   Composition of the synthetic HLLW DWK-HAWC-S/02

71

COMPOSITION OF SBR-S/05   (55 l - Mixture)

Comp.	Conc. (g/dm3)	Mass (g)	used Compound	Mass (g)	Oxide	Mass (g)	Conc. (g/dm3)
Rb	0,06	3,3	RbNO3	5,69	Rb2O	3,61	0,066
Sr	0,144	7,92	Sr(NO3)2	19,13	SrO	9,37	0,17
Y	0,079	4,37	Y(NO3)3 x 6 H2O	18,81	Y2O3	5,54	0,101
Zr	0,594	32,68	ZrO(NO3)2 x X H2O	98,08	ZrO2	44,14	0,803
			(45%ZrO2)				
Cd	0,011	0,63	Cd(NO3)2 x 4 H2O	1,72	CdO	0,72	0,013
Cs	0,633	34,8	CsNO3	51,03	Cs2O	36,89	0,671
Ba	0,241	13,24	Ba(NO3)2	25,19	BaO	14,78	0,269
La	0,207	11,37	La2O3	13,33	La2O3	13,33	0,242
Ce	0,41	22,55	Ce(NO3)3 x 6 H2O	69,89	CeO2	27,7	0,504
U	1,013	55,7	Ce(NO3)3 x 6 H2O	101,61	CeO2	40,28	0,732
Np	0,014	0,766	Ce(NO3)3 x 6 H2O	1,4	CeO2	0,55	0,01
Pu	0,196	10,8	Ce(NO3)3 x 6 H2O	19,22	CeO2	7,62	0,139
Am	0,601	33,07	Ce(NO3)3 x 6 H2O	59,09	CeO2	23,42	0,426
Pr	0,194	10,67	Pr(NO3)3 x 5 H2O	31,59	Pr6O11	12,9	0,235
Nd	0,653	35,93	Nd2O3 (95%)	44,12	Nd2O3	41,91	0,762
Pm	0,021	1,17	Nd2O3 (95)	1,41	Nd2O3	1,34	0,024
Sm	0,131	7,23	Sm(NO3)3 x 5 H2O	20,49	Sm2O3	8,38	0,152
Eu	0,012	0,66	Eu2O3	0,77	Eu2O3	0,77	0,014
Gd	0,008	0,44	Gd(NO3)3 x 5 H2O	1,21	Gd2O3	0,51	0,009
Tc	0,139	7,63	Mn(NO3)2 x 4 H2O	19,34	MnO	5,47	0,099
Ag	0,017	0,95	AgNO3	1,5	Ag2O	1,02	0,019
Ru	0,361	19,87	Ru(NO)(NO3)3-sol.	198,73	RuO2	26,11	0,476
			10 % Ru-content				
Rh	0,109	5,98	Rh(NO3)3 -solution	59,8	Rh2O3	7,37	0,134
			10 % Rh-content				
Pd	0,174	9,57	Pd(NO3)2 -solution	95,73	PdO	11,01	0,2
			10 % Pd-content				
Te	0,089	4,91	H6TeO6	8,84	TeO2	6,14	0,112
Sn	0,015	0,81	Sn	0,81	SnO2	1,03	0,019
Sb	0,004	0,22	Sb	0,22	Sb2O3	0,26	0,005
Mo	0,56	30,8	Na2MoO4 x 2 H2O	77,67	MoO3	46,21	0,84
HNO3	3 mol/l						
					Total	398,38	7,246

Table 5:   Composition of the synthetic HLLW SBR-S/05

(LEWC Reference Waste   Status: 17. August 1981 from Hahn-Meitner-Institut)

Comp.	Conc. (g/dm3)	Mass (g)	used Compound	Mass (g)	Oxide	Mass (g)	Conc. (g/dm3)
Rb	0,3	3	RbNO3	5,18	Rb2O	3,28	0,328
Sr	0,64	6,4	Sr(NO3)2	15,46	SrO	7,57	0,757
Y	0,57	5,7	Y(NO3)3 x 6 H2O	24,56	Y2O3	7,24	0,724
Zr	9,7	97	ZrO(NO3)2 x x H2O (45%ZrO2)	291,17	ZrO2	131,03	13,103
Ru	1,88	18,8	Ru(NO)(NO3)3 -sol. 20 % Ru-content	94	RuO2	24,75	2,475
Rh	0,36	3,6	Rh(NO3)3 -solution 10 % Rh-content	36	Rh2O3	4,44	0,444
Cd	0,09	0,9	Cd(NO3)2 x 4 H2O	2,2	CdO	0,92	0,092
Cs	1,98	19,8	CsNO3	29,04	Cs2O	20,99	2,099
Ba	1,44	14,4	Ba(NO3)2	27,4	BaO	16,08	1,608
La	1,18	11,8	La(NO3)3 x 6 H2O	36,78	La2O3	13,84	1,384
Ce	2,38	23,8	Ce(NO3)3 x 6 H2O	73,76	CeO2	29,24	2,924
U	1,44	14,4	Ce(NO3)3 x 6 H2O	26,27	CeO2	10,41	1,041
Pu	0,091	0,91	Ce(NO3)3 x 6 H2O	1,62	CeO2	0,64	0,064
Pr	1,1	11	Pr(NO3)3 x 5 H2O	32,55	Pr6O11	13,29	1,329
Nd	3,67	36,7	Nd(NO3)3 x 5 H2O	106,95	Nd2O3	42,81	4,281
Sm	0,84	8,4	Sm(NO3)3 x 5 H2O	23,82	Sm2O3	9,74	0,974
Eu	0,14	1,4	Eu2O3	1,62	Eu2O3	1,62	0,162
Gd	0,11	1,1	Gd(NO3)3 x 5 H2O	3,03	Gd2O3	1,27	0,127
Cr	3,12	31,2	Cr(NO3)3 x 9 H2O	240,11	Cr2O3	45,6	4,56
Mn	6,37	63,7	Mn(NO3)2 x 4 H2O	291,04	MnO	82,25	8,225
Tc	0,78	7,8	Mn(NO3)2 x 4 H2O	19,78	MnO	5,59	0,559
Fe	19,4	194	Fe(NO3)3 x 9 H2O	1403,4	Fe2O3	277,37	27,737
Zn	0,21	2,1	Zn(NO3)2 x 4 H2O	8,4	ZnO	2,61	0,261
Ni	5,51	55,1	Ni(NO3)3 x 6 H2O	272,97	NiO	70,12	7,012
PO4	0,14	1,4	H3PO4 (85%)	----	P2O5	---	--
SO4	10,71	107,1	H2SO4	----	SO3	---	--
F	12,38	123,8	HF (40%)	----		---	--
Al	8,1	81	Al(NO3)3 x 9 H2O	1126,17	Al2O3	153,05	15,305
Pd	1,28	12,8	Pd(NO3)2 -solution 10 % Pd-content	128	PdO	14,72	1,472
Ag	0,05	0,5	AgNO3	0,79	Ag2O	0,54	0,054
Na	40,4	389,29	NaNO3	1439,23	Na2O	524,75	52,475
Te	0,53	5,3	H6TeO6	9,54	TeO2	6,63	0,663
Sn	0,05	0,5	Sn	0,5	SnO2	0,63	0,063
Sb	0,015	0,15	Sb	0,15	Sb2O3	0,18	0,018
Mo	3,07	30,7	Na2MoO4 x 2 H2O	77,42	MoO3	46,06	4,606
Na		14,71			Na2O	19,83	1,983
HNO3	2,5 m/l						
						Total	1589,09

Table 6:   Composition of the synthetic HLLW
LEWC-S/01

73

Fig. 1: Denitration step of the FIPS-II equipment

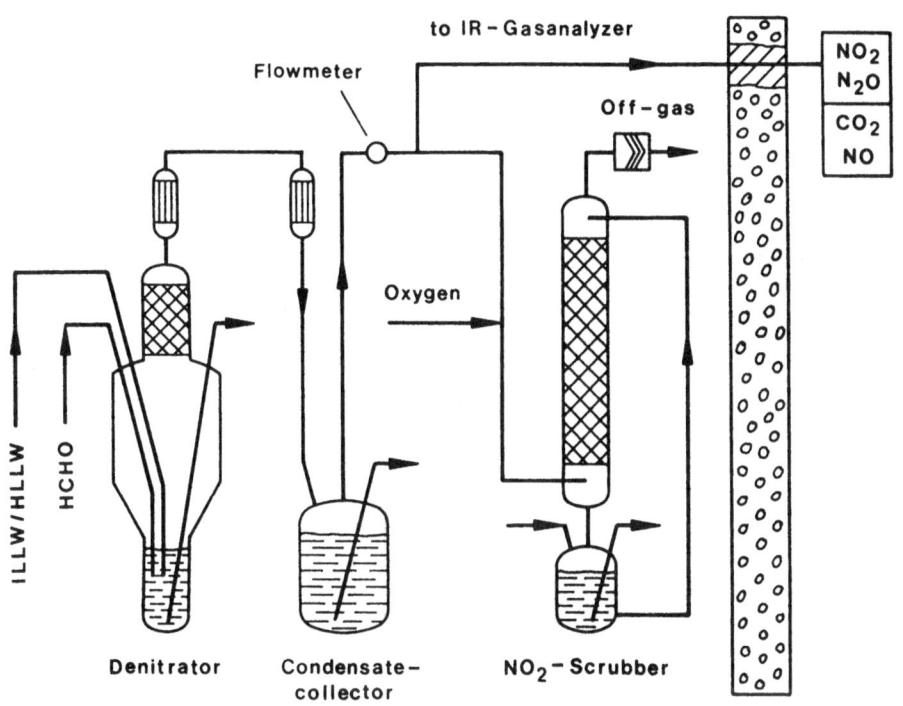

Fig. 2:   Simplified flow sheet of FIPS-II denitration

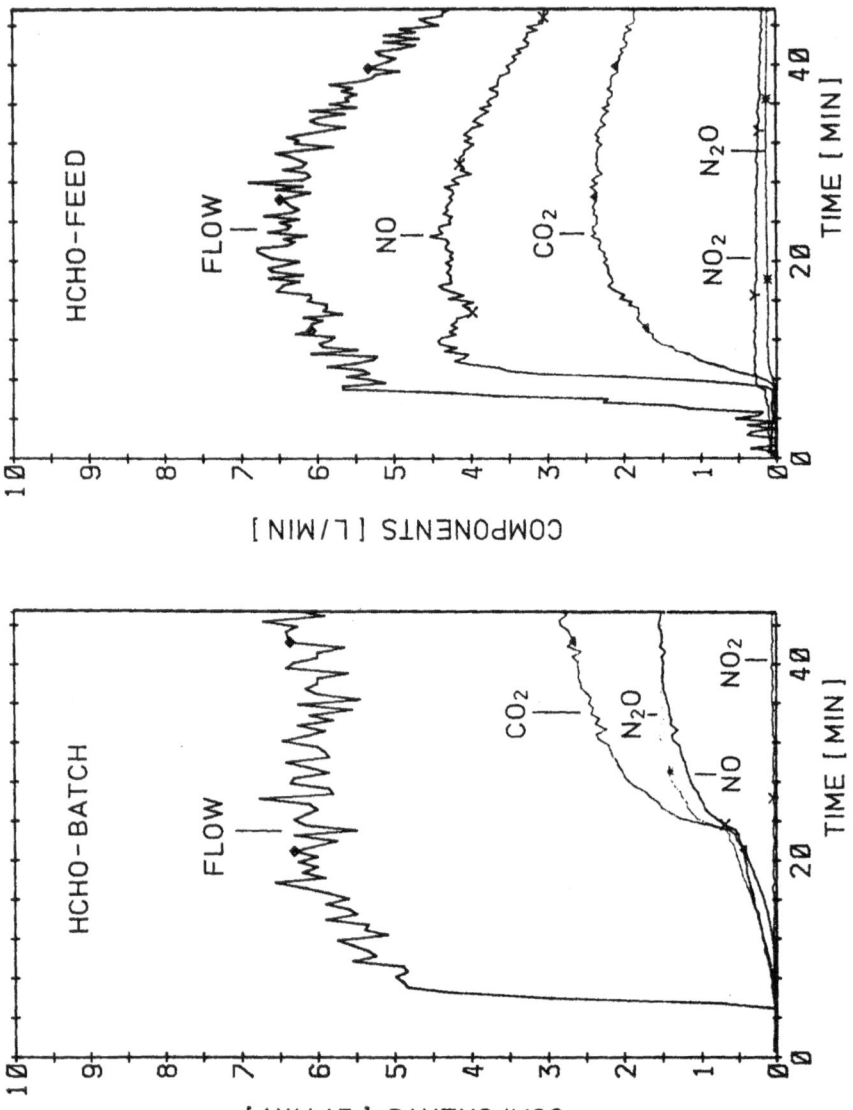

Fig. 3: Comparison between the gas evolution rates during denitration

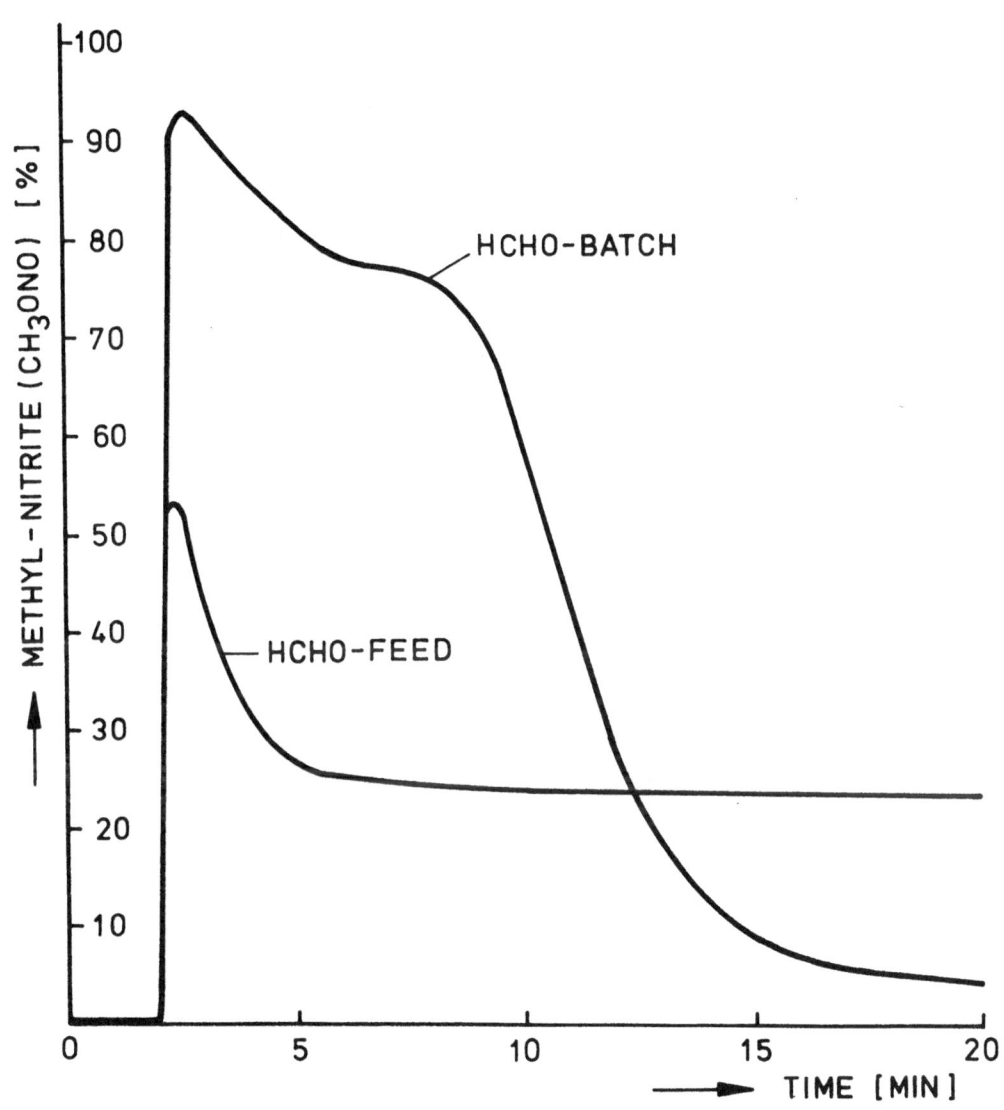

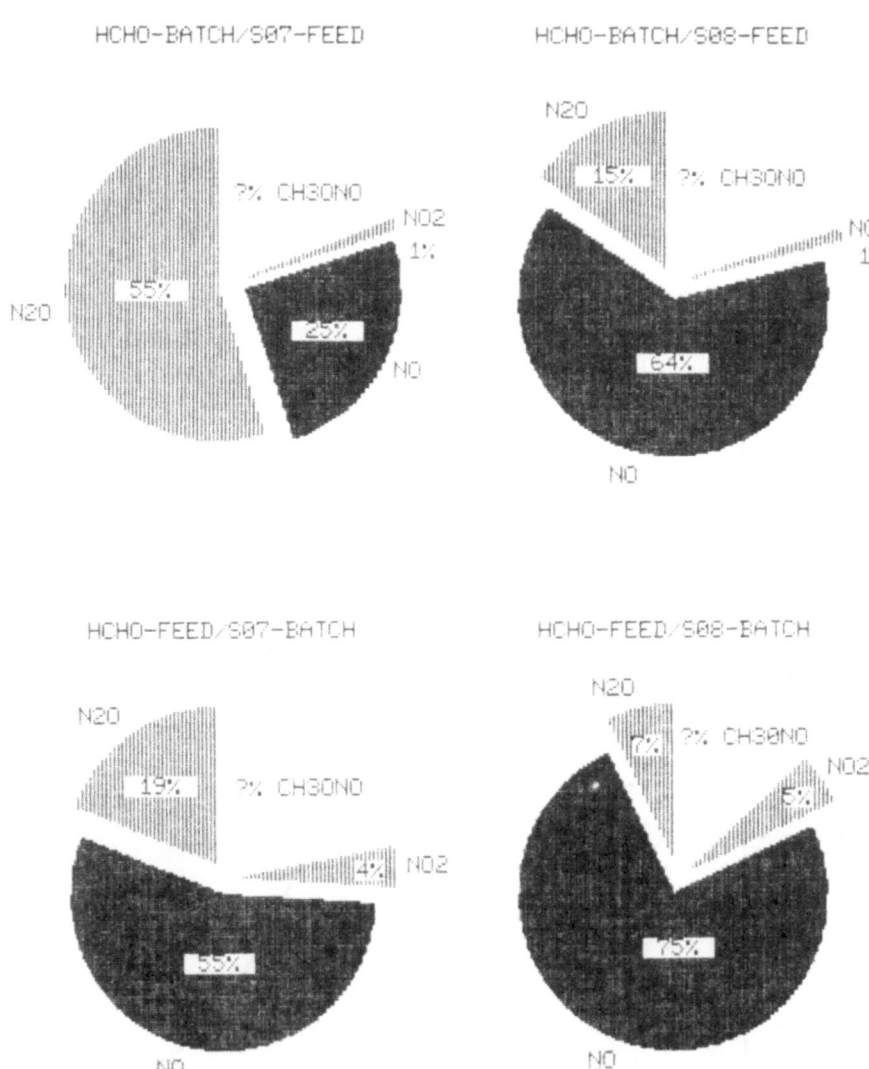

Fig. 5: Balance of the conversion of nitric acid to nitric oxides for the denitration of ILLW

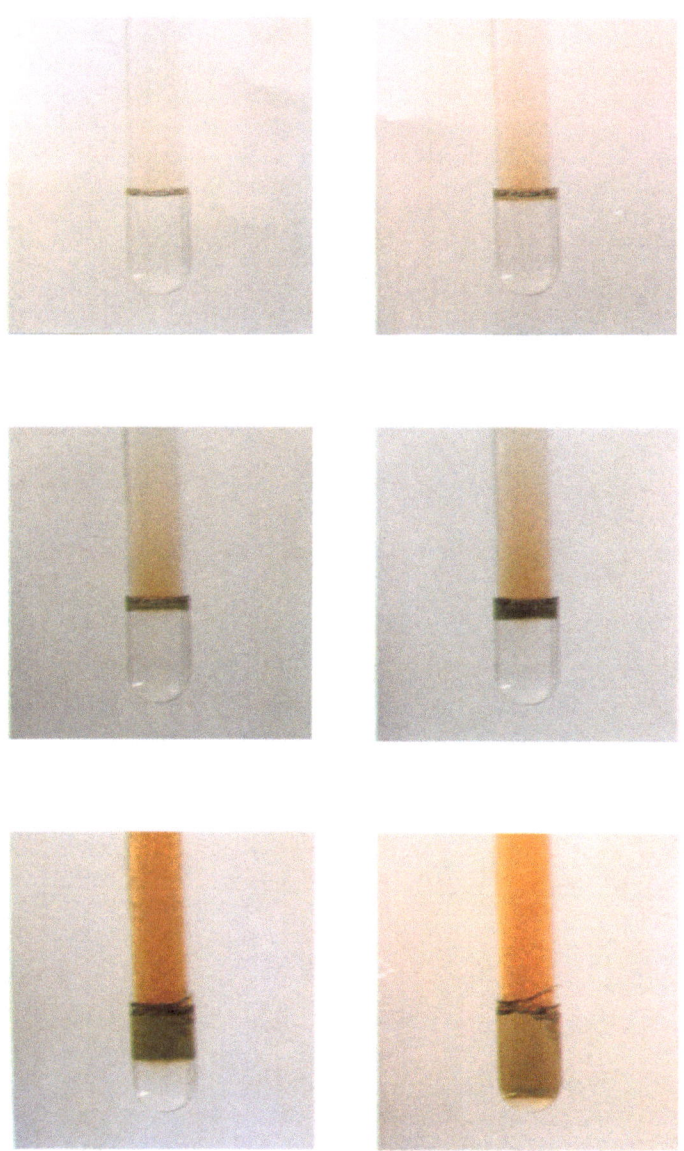

Fig. 10: Autocatalytic reaction in a mixture of unstabilized HCHO and HNO_3

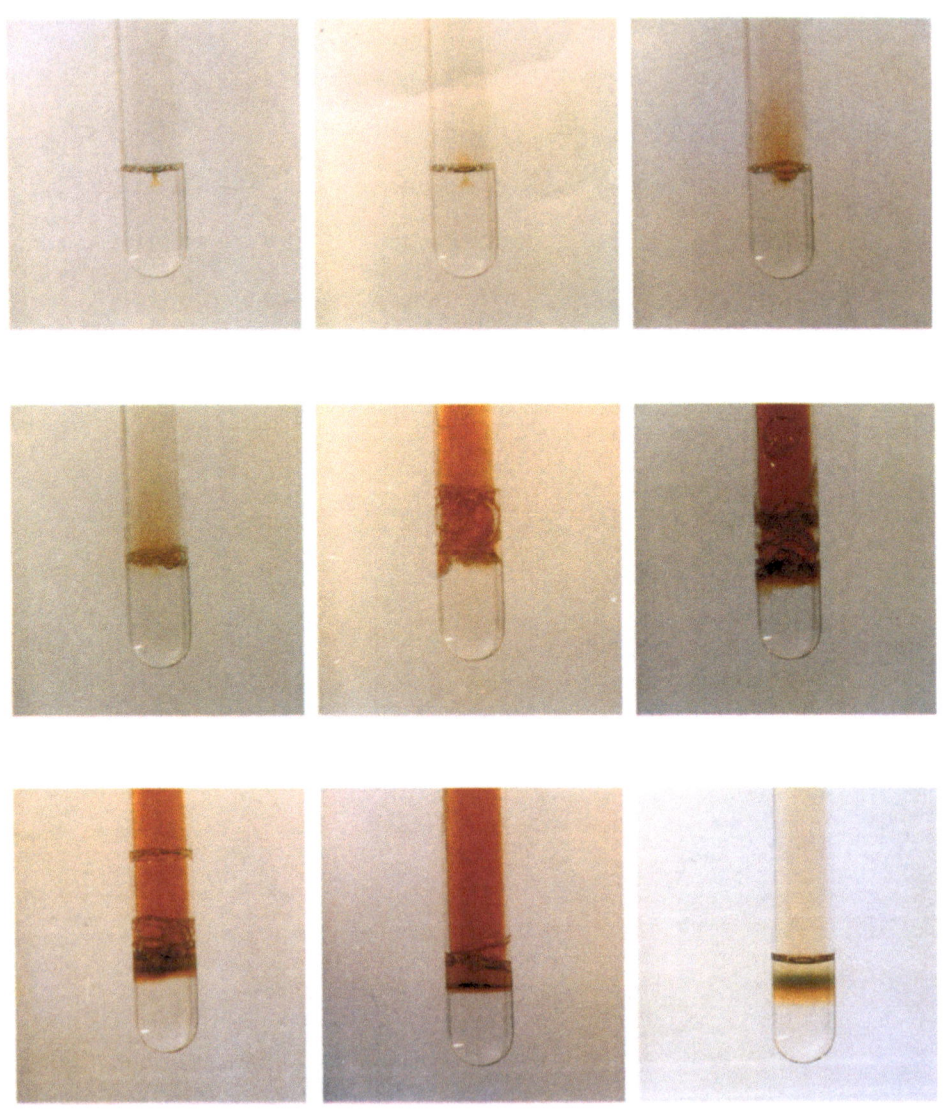

Fig. 11: Autocatalytic reaction between HNO₃ and a layer
of unstabilized HCHO

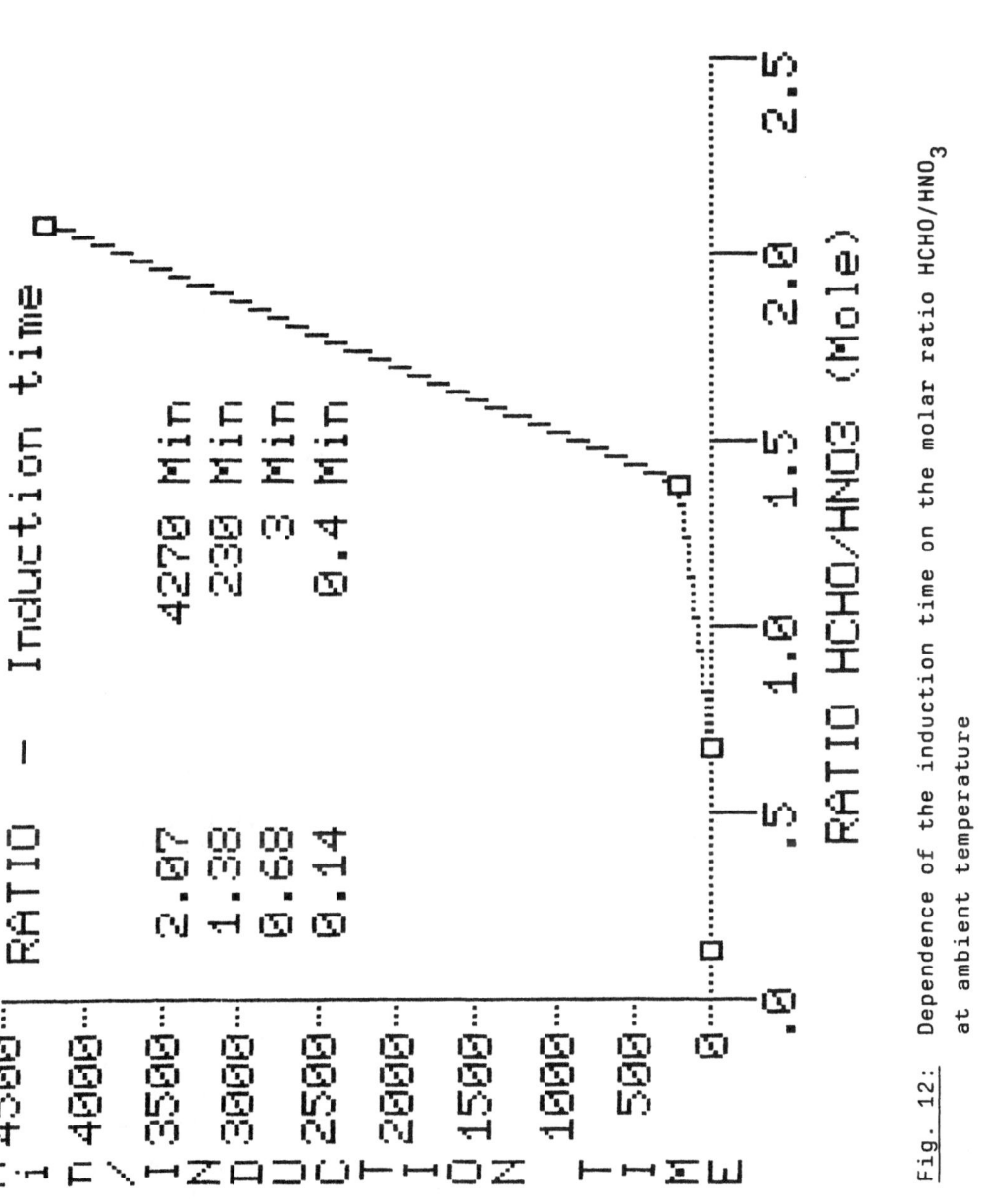

Fig. 12: Dependence of the induction time on the molar ratio HCHO/HNO₃ at ambient temperature

DENITRATION OF REPROCESSING CONCENTRATE BY MEANS OF HCOOH

M. Kelm, B. Oser, S. Drobnik
Kernforschungszentrum Karlsruhe GmbH
Institut für Nukleare Entsorgungstechnik

Summary

A batch process has been developed to destroy nitric acid present in waste solutions from reprocessing plants by formic acid. The main reaction products N_2O and CO_2 may be released into the atmosphere.

The process was subjected to testing on lab scale involving simulated and real medium-level waste concentrate, on semitechnical scale involving 50 l, and on pilot scale involving 200 l batch volumes of simulated medium-level waste solutions. In the denitration runs between 81 and 99 % of the nitric acid could be destroyed. The final nitric acid concentration ranged from 0.2 to 0.02 mol/l. The aerosole drag-out with the reaction gases lies between 1 and 5 ppm related to the reaction vessel contents.

The amount of formic acid dragged out ranges from 0.2 to 0.3 %. The formic acid dragged out can be conveniently separated in quantitative terms in a scrubbing column.

Corrosion tests performed on material specimens and on the reaction vessel itself have shown that Incoloy 825 is a suitable construction material for the denitration vessel.

The explosion limits and pressures of formic acid vapour/water vapour/air mixtures were determined. The presence of an explosive mixture can be excluded downstream of the condenser.

A restart of the reaction after extended periods of interruption in metering which are necessary, e.g., in case of technical disturbances, was tested on the pilot facility under various conditions and does not cause problems. The associated induction period can be calculated in advance with sufficient accuracy using a mathematical approach developed from laboratory scale experiments. If the solution is cooled down immediately when metering is interrupted, the reaction restarts already during reheating, even before starting to meter the feed solution.

1. INTRODUCTION

Denitration reactions involving formic acid have been described in the literature already since 1954 /1,2,3/. Using these techniques the nitric acid waste solutions were fed into the reaction vessel and the reducing agent was slowly metered in. This procedure gives high NO_2 and NO contents in the reaction gases.

In the process version developed at KfK /4/ there is acted vice versa. Formic acid is fed into the reaction vessel, heated to boiling, and the nitric acid waste solution is subsequently metered in. If the reaction proceeds in this way, HNO_3 is always in a reducing environment so that the reaction gives mainly N_2O. Only towards the end of the metering phase low amounts of NO are formed. The advantage offered by this process version is that a high percentage of the nitric acid is converted into N_2O and may be discharged whereas practically all NO and NO_2 has to be recovered as nitric acid in an off-gas stripping if the process is run in the manner described first.

2. EXPERIMENTAL

The process of denitration was tested with simulated medium-level waste solutions on lab, semitechnical and pilot scale and with true MLW concentrates on lab scale. The reaction was performed batchwise. Using simulated solutions the batch volumes were 1 l on lab scale, 50 l on semitechnical scale, and 200 l in pilot scale. True, active waste concentrates were denitrated in 200 ml batches.

The process takes the following course:

- The stoichiometric quantity of concentrated formic acid (according to Eq.(1)) is fed into the reaction vessel and heated to boiling under reflux.

- Depending on the nitric acid content the waste solution is fed within 1.5 to 4 hours. During this process mainly carbon dioxide and dinitrogen monoxide are formed (Eq.(1)).

$$2 \; HNO_3 + 4 \; HCOOH \rightarrow N_2O + 4 \; CO_2 + 5 \; H_2O \qquad (1)$$

$$2 \; HNO_3 + 3 \; HCOOH \rightarrow 2 \; NO + 3 \; CO_2 + 4 \; H_2O \qquad (2)$$

$$2 \; HNO_3 + 5 \; HCOOH \rightarrow N_2 + 5 \; CO_2 + 6 \; H_2O \qquad (3)$$

- The denitration reaction starts after an induction period of a few seconds and then proceeds calmly controlled by the metering rate.

- In side reactions according to Eqs. (2) and (3) some percent of NO and a little N_2 are formed additionnally.

- To have the reaction completed the solution is kept boiling for another two hours after metering has been finished.

- The remaining amount of formic acid is destroyed by addition of hydrogen peroxide.

The laboratory-scale tests were performed in a glass flask with a reflux condenser attached to it; the pilot tests were performed in a coolable and heatable stainless steel tank of 1.6 m^3 volume (Fig.1). In this system 40 m^3 simulated waste solution in total were denitrated in about 200 single tests. In 16 tests 3.2 l true MLW concentrates of various origin were denitrated. The simulated medium-level waste solutions contained up to 2.5 mol/l sodium nitrate and up to 4 mol/l nitric acid. Heavy metals were simulated by 0.06 mol/l iron nitrate.

3. RESULTS

In all denitration tests the turnover, the residual acid content, the rate of gas evolution, and the compositions of the reaction products were determined.

Figure 2 shows a nitrogen balance of denitration of simulated and true MLW concentrates. Between 81 and 99 % of the nitric acid have been destroyed. Up to 83 % of the nitric acid were reduced into N_2O, 15 to 20 % into nitrogen oxide, and about 3 % into nitrogen. The non-toxic gases contain 60 to 85 % of the nitrogen added. The portion of nitrogen reduced to NO increases with increasing metering rate, growing nitric acid and heavy metal concentrations in the feed solution and with the use of a substoichiometric amount of formic acid.

Figure 3 shows the typical course of the rate of gas evolution, the N_2O content and the NO content in the reaction gas at a constant metering rate of waste solution. The rate of NO evolution attains its maximum at the end of metering.

After denitration the nitric acid content was between 0.2 and 0.02 mol/l. The lowest nitric acid concentrations were attained with solutions containing sodium nitrate. An addition of chemical decontaminants such as citric acid, oxalic acid and tensides has no negative effect on the reaction. No noticeable problems were encountered by foam formation.

The increase in volume during denitration ranges from 2 to 7 %. Oxidation with H_2O_2 (30 wt.%) leads to a further dilution by 5 to 7 %.

The tests involving true MLW concentrates yielded results similar to those obtained with simulated solutions (Fig. 2). During denitration 2 to 6 g/l were precipitated. Their main constituents were almost all of the uranium present in the MLW and a considerable fraction of the iron as well as up to 7 % of the gross beta activity inserted. No ruthenium precipitation was observed.

4. SAFETY ASPECTS TO BE CONSIDERED DURING DENITRATION

4.1 Aerosol drag-out

As the denitration reaction is accompanied by a strong evolution of gas, the drag-out of radionuclides as aerosols has to be expected when radioactive solutions are handled. The respective quantitative measurements were performed on lab scale with simulated and true medium-level waste solutions and on pilot scale with simulated solutions. In the lab scale tests the aerosols were removed on a series of two filters. In the inactive tests sodium and rubidium, respectively, were used as tracer elements for the quantitative evaluation. During denitration of true medium-level wastes the volatility of the elements Ce, Ru, Cs, Sr and Sb was investigated separately by an activity measurement. An aerosol drag-out of ≤ 5 ppm, related to the composition of the feed solution was found, which hardly varies from element to element. The tests performed at the pilot facility with a scintillation particle counter and sodium as the tracer element yielded slightly more detailed findings. Accordingly, aerosol drag-out varies but insignificantly with the metering rate and increases during the metering phase from 0.15 ppm h^{-1} to 0.4 ppm h^{-1} related to the total sodium content in the reaction vessel. At the end of metering aerosol drag-out rapidly approaches zero.

4.2 Volatility of the formic acid

Depending on the respective concentration and the partial pressure of the formic acid in the condenser, formic acid is dragged out together with the off-gas. At 303 K the maximum formic acid concentration in the off-gas downstream of the condenser (Fig. 1) was 2.8 mol/m^3 at the beginning of metering and it quickly dropped to 0.01 mol/m^3 during the first third of the metering time. Downstream of a water operated scrubbing column formic acid has never been found in the off-gas. The total formic acid drag-out, measured downstream of the condenser, is 0.2 to 0.3 % of the amount added in one denitration batch.

4.3 Corrosion

In the course of development on the process of denitration the use of concentrated formic acid seemed to be disadvantageous due to corrosion phenomena. Comparative studies on 27 kinds of stainless steel yielded very great differences in the corrosion rates which were particularly high in aerated formic acid. Incoloy 825 was chosen as the most resistant material, which, under these conditions, showed a corrosion rate of 0.5 mm a^{-1}. Detailed investigations with this material have shown that in load cycles characteristic for denitration the corrosion rates attain only 0.014 mm a^{-1}. These results were verified in extensive tests relying on radionuclide technique and performed at the Laboratory for Isotope Technology/KfK. With highly conservative assumptions on the cycle durations in denitration - e.g., regular extended boiling of concentrated formic acid before metering - about 10 times the previous corrosion rates are obtained, namely 0.16 mm a^{-1} with 400 batches. This value is equally applicable to welded seams.

Due to these results, the denitration vessel and the condenser of the pilot scale equipment were made from Incoloy 825. During manufacture the supplier produced work specimens with welded seams which were used at KfK for stress corrosion examinations performed under realistic conditions. After 20 denitration cycles intergranular corrosion phenomena in excess of a minor attack to 15 µm depth which had been the result of improper pickling were neither found on the specimens stored in the liquid phase nor on the specimens from the gas phase. Reference specimens with milled surfaces did not show any attack under identical conditions, the same as specimens subjected to an aggravated Strauß test according to DIN 50 914 with three times the volume of sulfuric acid.

The examination of the internal surfaces of the denitrating vessel after about 100 batch tests using plastic impression and dye penetrating tests and by means of optical inspection under the microscope did not provide an indication of intergranular corrosion.

4.4 Induction period

Denitration is characterized by an induction period depending on the temperature and the concentrations prevailing at the beginning of the reaction. The induction period is an important parameter under safety aspects because during this period nitric acid may accumulate in the formic acid fed into the reaction vessel.

Under these conditions a delayed start of reaction is paralleled by a sudden high release of heat and gas and the resulting problems such as overloading of the reaction gas cooler, foaming over, and pressure build up. Because of this safety relevance efforts were undertaken, in cooperation with the Technical University of Hannover /5/ to describe in mathematical terms the induction periods found in the experiments by their dependence on temperature and concentration (Eq.(4)). In the literature /2,3,/ the delay in initiation of the denitration reaction is linked to the occurrence of autocatalytically formed nitrite ions. Measurements by ion chromatography have shown that during the reaction nitrite ions are actually formed at concentrations up to 0.1 mol/l in the reaction pool. If we assume that a given minimum concentration of this intermediate product must exist at the beginning of the reaction, its rate of formation is inversely proportional to the induction period measured. Therefore, the values measured for the induction period should be describable by a reaction rate law in a broad range of temperatures and concentrations (Fig. 4).

$$\frac{1}{\tau} = 2.8 \times 10^{29} \cdot c_A^{\ 3} \, c_N \, \exp\left(-\frac{26040}{T}\right) \qquad (4)$$

where

c_A, c_N = concentrations (HCOOH and HNO_3, resp.) in mol/l,

$\quad T$ = temperature in K,

$\quad \tau$ = induction period in minutes.

At the boiling temperature and with concentrated formic acid fed into the reaction vessel an induction period of 5 s can be calculated with this equation. Actually, both on lab scale and on pilot scale only an induction period of a few seconds duration was observed at the beginning of the reaction.

4.4.1 Induction period in case of restart of the reaction after interruption of metering

Technical defects on the system as well as troubles may cause an interruption of feed metering of several hours duration. In order to avoid faulty batches the question must be answered how to restart the reaction after an extended interruption of metering. On account of the lower concentration of the reactants the induction periods were longer.

Experiments to determine the induction periods at the moment of restart of the reaction were performed on the pilot facility. After one hour of interruption of metering the induction period increased from 1.2 minutes at 7 mol/l formic acid concentration to 2.2 minutes at 1.3 mol/l formic acid concentration in the reaction vessel. When metering was interrupted for 17 hours while the solution continued to boil, the induction period with a formic acid concentration of 1.3 mol/l was 3.5 minutes. Due to heating during long feeding interruption the intermediate compound necessary for restarting the reaction is obviously destroyed. Despite the relatively long induction period observed in this case, the reaction starts slowly also after interruption of metering and without violent gas evolution.

If the reaction solution is cooled immediately upon interruption of metering, the reaction restarts already during reheating and even before metering starts again (Fig. 5).

4.5 Possible endangering of the denitration plant by gas explosions

The generation of an explosion presupposes the combined effect of three conditions:

a) the presence of an oxidable matter, in this case formic acid

b) the presence of an ignition source

c) its enrichment with air or other oxidants, to form an explosive mixture.

If only one of these three premises is missing, no explosion can occur.

To prevent an explosive mixture, the knowledge of the explosion limits of the gas mixtures is involved of important safety relevance. In cooperation with the Fraunhofer-Institut for fuels and explosives, the explosion limits and explosion pressures of the system formic acid vapour/water vapour/air have been determined experimentally (Fig. VI).

The maximum explosion pressure is approx. 3.7 bar, related to a mixture composition of 32 vol % formic acid vapour, and 67.7 vol % air.

Due to the analysis of formic acid in the off-gas system, there is no explosive mixture present downstream the condenser. This has also been confirmed by an explosion monitor. At starting of metering, the maximum deflection of this monitor was at approx. 10 %. The full scale range of the monitor was adjusted with 2 % H_2 in air (equivalent to 50 % of the lower explosion limit).

In the reaction vessel, the formic acid is diluted rapidly by water after starting metering. When the formic acid content in the liquid phase has fallen below 79 % weight percent formic acid, here, too, no explosive mixture can be generated any more.

The ignition temperature of the formic acid lies at 520¬C. As the vessel is heated with steam (3 bar, 408 K), the heating system is excluded as an ignition source.

All results of our investigations regarding denitration of MLW concentrate incl. the investigations of kinetic reaction have been summarized in a KfK report /4/.

REFERENCES

/1/ J.V.Longstaff; K.Singer
    The Kinetics of Oxidation by Nitrous Acid and Nitric Acid,
    Part. II, Oxidation of Formic Acid in Aqueous Nitric Acid
    J. Chem. Soc. (1954), 2604.
/2/ T.V.Healy
    The Reaction of Nitric Acid with Formaldehyde and with Formic Acid
    and its Application to the Removal of Nitric Acid from Mixtures
    J. Appl. Chem. 8 (1958) 553.
/3/ R.F.Bradley; C.B. Goodlett
    Denitration of Nitric Acid Solutions by Formic Acid
    USA EC Report DP-1299
    Savannah River Laboratory, E.I. du Pont de Nemours and Company
    Aiken S.C. 29801 (1972).
/4/ M.Kelm; B.Oser; S.Drobnik; W.Bähr
    Die chemische Denitrierung von salpetersauren schwach- und
    mittelaktiven Abfallösungen mit Ameisensäure
    KfK 3948, October 1985
/5/ K.Holze; H.D.Finke; M.Kelm; W.D.Deckwer
    Reaction Model for Denitration with Formic Acid of Waste Effluents
    from Nuclear Fuel Reprocessing Plants
    Ger. Chem. Eng. 2 (1979) 361-371.

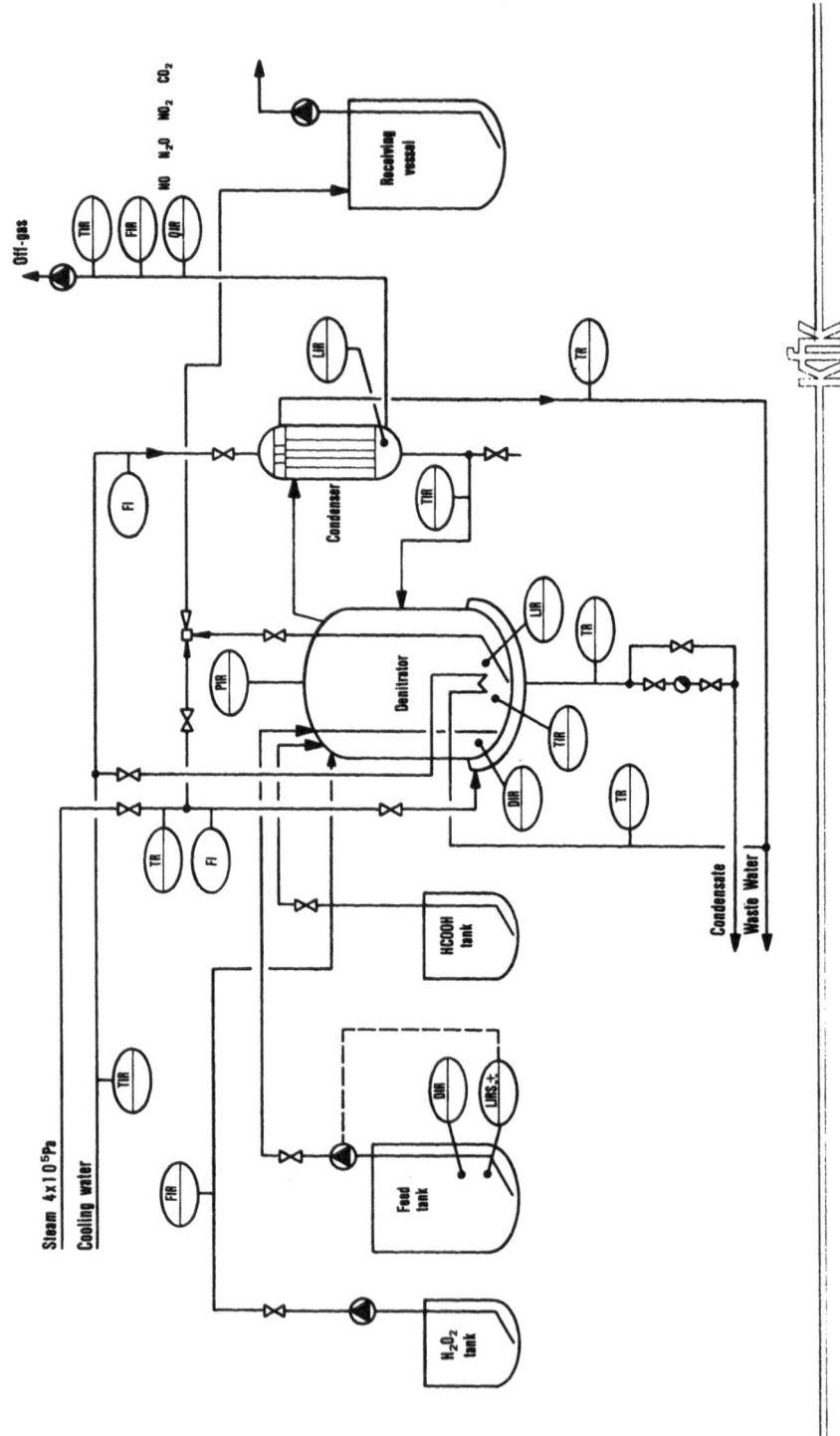

Fig. 1: Simplified scheme of the pilot scale denitration facility

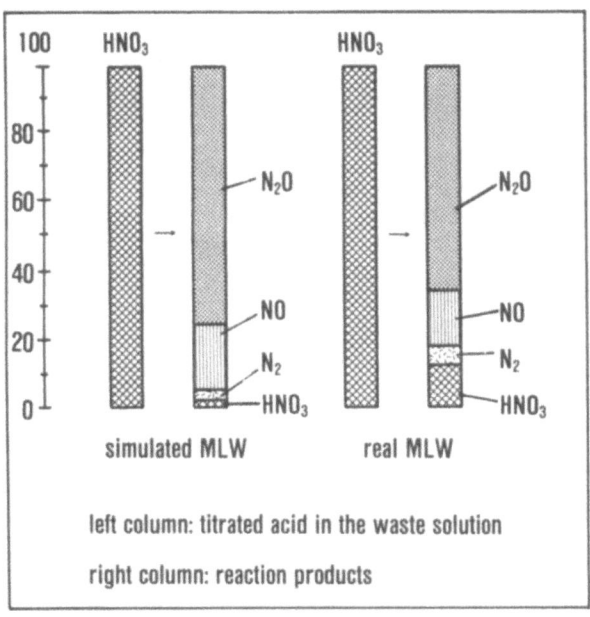

Fig. 2: Nitrogen balance from various denitration runs

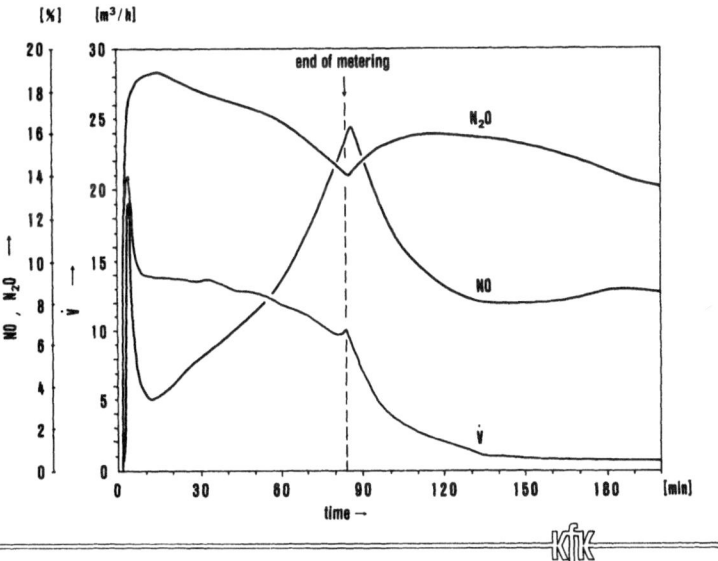

Fig. 3: Gas evolution rate, NO and N₂O proportion in the reaction gas depending on time

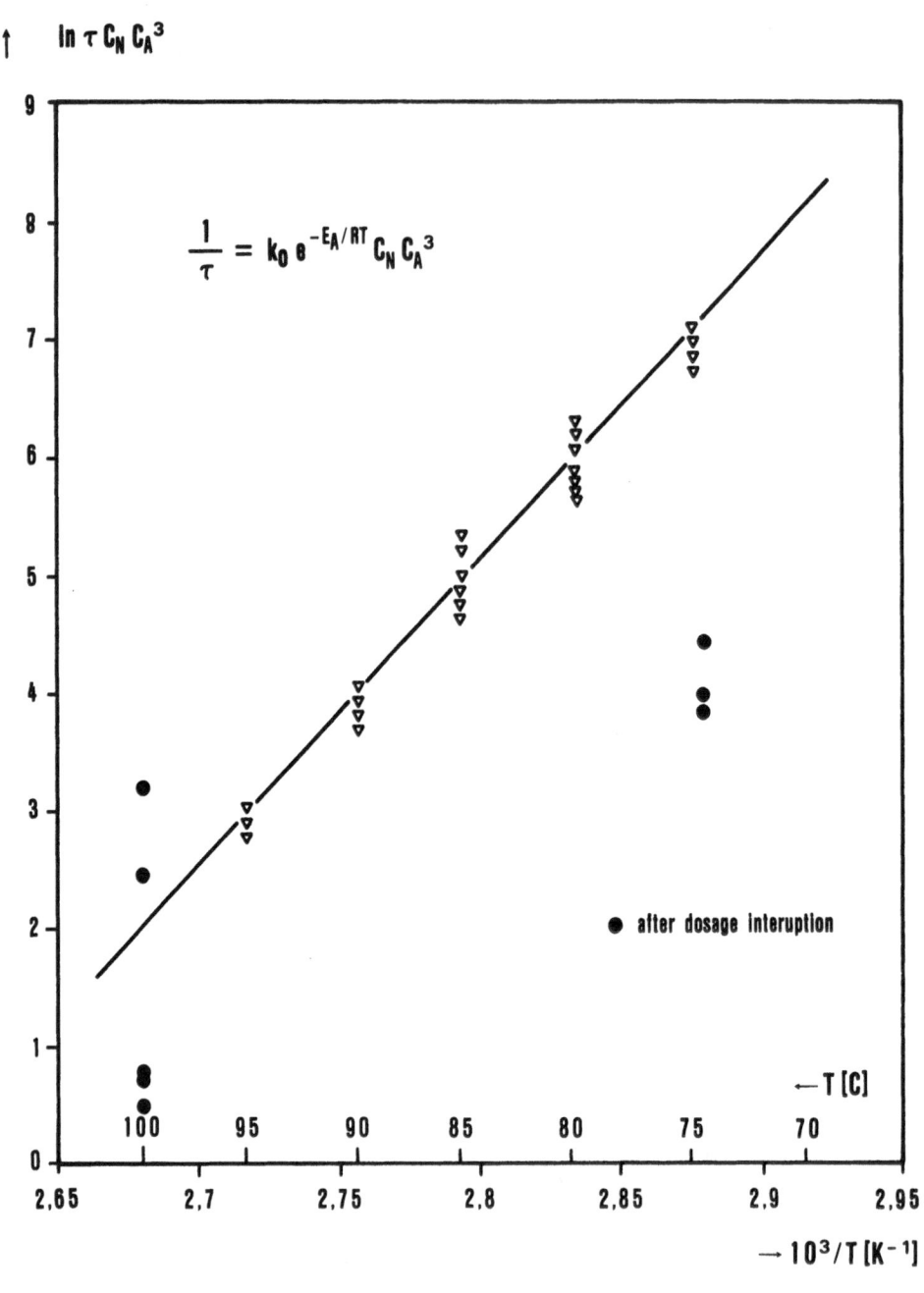

Fig. 4: Induction period on the denitration reaction

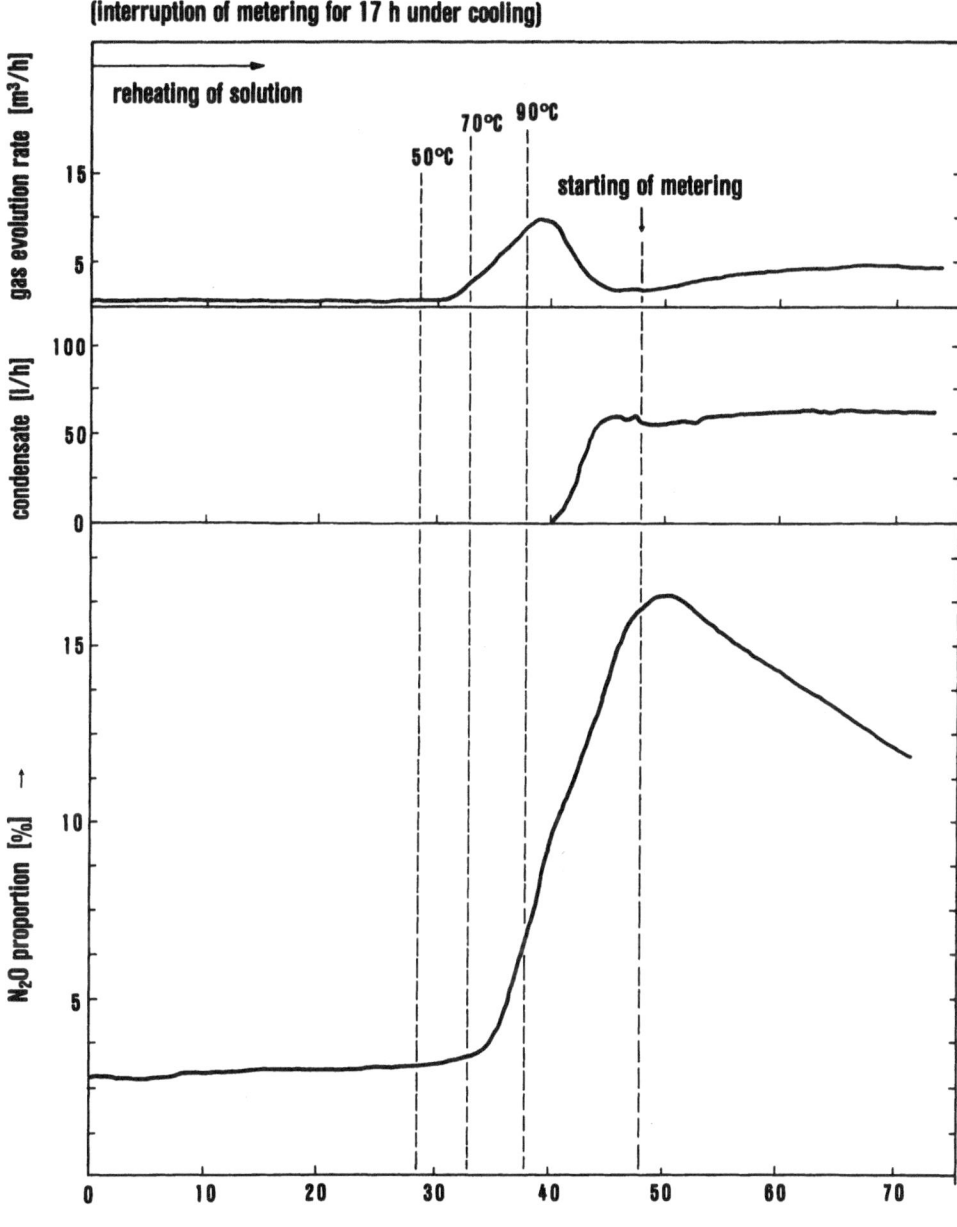

Fig. 5: Gas evolution rate, condensate reflux and N_2O proportion during reheating and starting of reaction

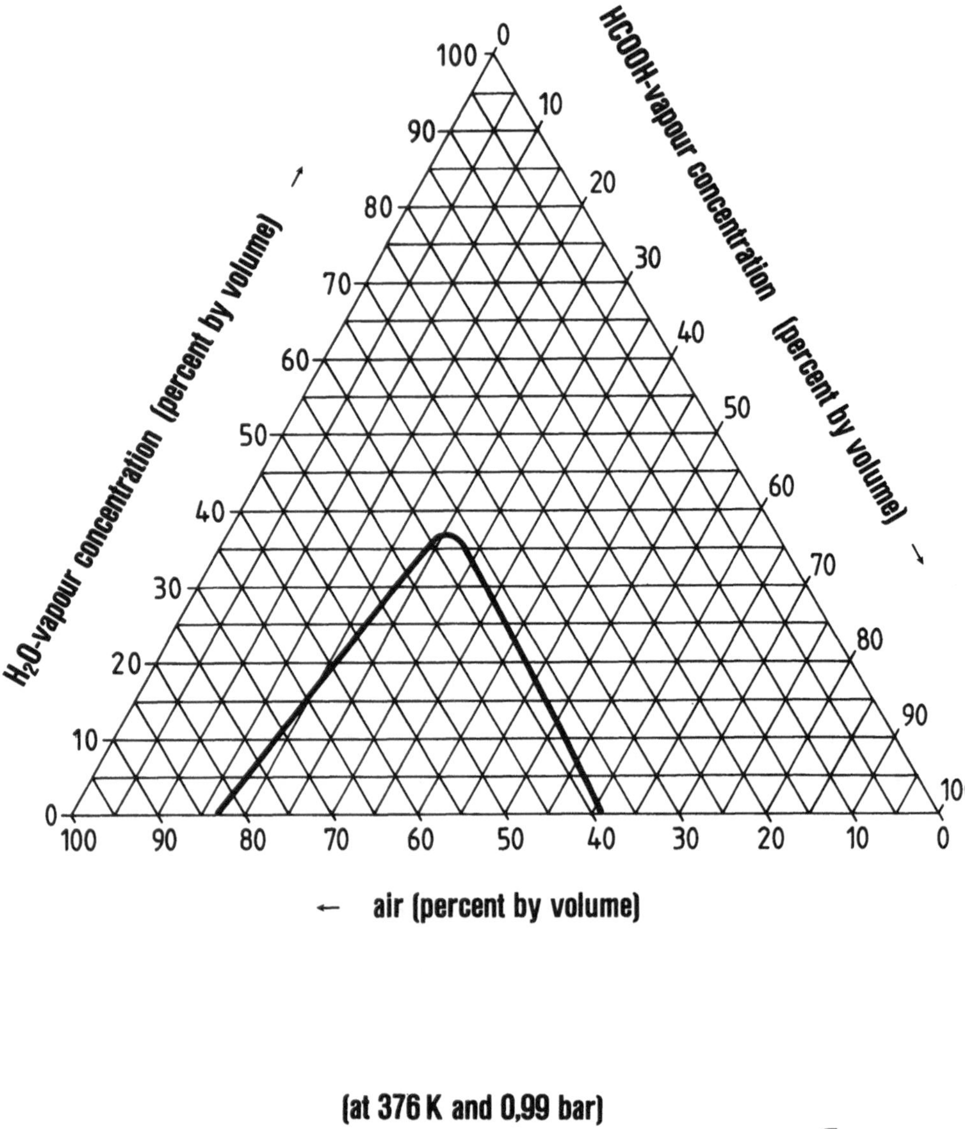

100 ⌐ 0

HCOOH-vapour concentration (percent by volume) →

H₂O-vapour concentration (percent by volume) ↑

air (percent by volume) ←

(at 376 K and 0,99 bar)

Fig. 6: Explosion limits of system HCOOH-vapour/H₂O-vapour/air

APPLICATION OF DENITRATION/OXALATE PRECIPITATION IN THE PETRA HOT-CELL FACILITY

G. VASSALLO*, H. DWORSCHAK*, F. SEGANTINI**, A. FACCHINI**

*C.E.C.-J.R.C. Ispra Establishment, **Polytechnic of Milan

Summary

The PETRA hot-cell facility has been developed to study HAW conditioning alternatives. The first strategy to be assessed includes the denitration of high active waste (HAW) and the contemporaneous oxalate precipitation of the rare earths and actinides. A "cold" pilot plant was constructed to establish design and operating conditions which could be successfully applied within the facility. All experiments were based on feeding simulated HAW into the denitrator containing formic acid. The problem of induction period may be overcome by feeding the HAW locally, rather than well distributed, into the reactor. However, the presence of strong nitrous acid scavengers must be eliminated as these could block the denitration reaction. In this context, iron plays an intermediate role, partially protecting the oxalic acid from nitrous acid, while allowing the formic/nitric acid destruction to proceed. Off-gas and pressure measurements reliably detect the commencement and continuation of the process while electrical conductivity indicates the prevailing nitric acid concentration. After HAW addition, a reflux period of several hours is necessary to attain sufficiently low nitric/formic acid concentrations. Following supernatant removal, the recovery and destruction of the oxalate precipitate entail its dissolution in dilute nitric acid and subsequent concentration.

1.0 INTRODUCTION

The need to reduce the nitric acid concentration in waste solutions can be effectively achieved using organic reagents such as formic acid. With high active waste (HAW), this operation can be modified to include oxalic acid in order to obtain a substancial partitioning of actinides and rare-earths from fission products; the former group precipitates as oxalate salts while most other cations remain in solution (1). A pilot plant based on this process was constructed for denitrating the equivalent inactive HAW that would arise from the reprocessing of 6.8kg of spent UO_2 fuel. This is similar in size to the fuel batches which will be handled in the new hot-cell facility, PETRA, at the J.R.C., Ispra Establishment (2).

Although several operating modes exist, only one, namely the addition of simulated HAW to a very hot or boiling solution of formic/oxalic acid was investigated. In general, $31.5dm^3$ of simulated HAW were treated with several litres of pure formic acid.

2.0 PILOT_PLANT_DESCRIPTION

The pilot plant was primarily composed of five glass vessels: simulated HAW make-up vessel, HAW feed vessel, denitrator/oxalate precipitator, oxalate concentrator and wash column. These were placed in two stainless steel catch trays and supported, when necessary, with aluminium scaffolding, Figure 1.

The denitrator, a 100dm³ capacity vessel, was heated using a mantle equipped with three 2kW horizontal resistance bands. Apart from general process tubing and instrumentation, the inside of the denitrator also housed a cooling coil and sampling system. Immediately above was a water cooled condenser offering 1.5m² of heat transfer area. Non-condensable gas passed through the condenser to an 800mm high columnar scrubber. This was filled with glass helices through which a solution of nitric acid and hydrogen peroxide was recirculated using a piston pump. The resulting off-gas was then directed to a small stack situated above the laboratory.

The 50dm³ simulated HAW make-up and 10dm³ oxalate concentrator vessels were also fitted with heating mantles. Their contents together with most other process liquids were transferred by suction with a vacuum pump. However, for dosing purposes, the HAW feed vessel was connected to the denitrator with a 0-40dm³/h bellows pump. The process inter-connecting tubing was generally stainless steel as were most valves. Many of the valves could be activated remotely thereby rendering the plant semi-automatic. A schematic representation of the process circuit is given in Figure 2.

The whole plant was equipped with a wide range of instrumentation for either proving applicability, monitoring the state of the denitration process or safety purposes. The majority of measurements was displayed on a conventional panel; some measurements were also sent to a data acquisition service.

3.0 OPERATING_PROCEDURE

Several types of simulated HAW were employed. The most simple, consisting of nitric acid, was prepared directly in the HAW feed vessel. Noble metals and many different rare earths, representing a total of 26 elements, were included in the most complex solution. In this case the HAW was produced by adding pure metal, oxide or nitrate salts to nitric acid and then boiling under total reflux for several hours in the make-up vessel. Complete dissolution of the salts was quickly attained except in the case of molybdenum and zirconium; even after many hours much of these elements could still be found as a glutinous solid. When cool, the simulated HAW was filtered and transferred to the feed vessel.

Only formic acid, added directly to the denitrator, was utilised in certain experiments. In others, crystalline oxalic acid was placed in the oxalate concentrator. Formic acid was then added and the mixture heated until the oxalic acid dissolved. The still warm mixture was then transferred to the denitrator. The inherent problem of inhaling either formic acid fumes or fine particles of oxalic acid was always evident during organic reagent preparation. During this period, the simulated HAW was recirculated through the HAW feed vessel to homogenise the solution. The organic reagents were then quickly heated in the denitrator. When they had attained the required temperature, 101-105°C, simulated HAW was added at a nominal flowrate of 20dm³/h. This value reduced the feeding period to a reasonable level while giving enough time to intercede if necessary.

Figure 1. DENITRATION PILOT PLANT.

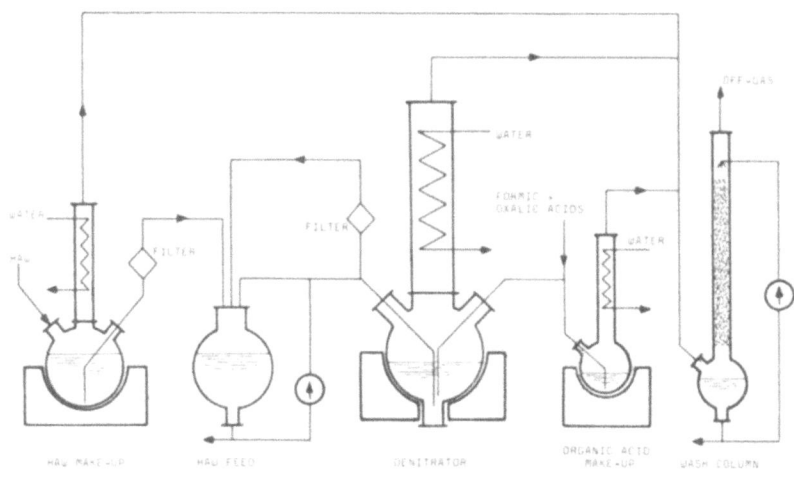

Figure 2. SCHEMATIC DIAGRAM OF FLOWSHEET.

Approximately 1.75h were required to feed the HAW. The dosing pump was then flushed with a litre of dilute nitric acid to ensure that all of the original feed reached the denitrator, thereby preventing post-precipitation of any salts in the process tubing. The denitrated solution was then totally refluxed for a minimum period of 1.5h by which time the off-gas production was generally low. Samples of liquid could be taken throughout this operation. The sampling line was normally purged with argon. However, at programmed intervals, the flow of gas was stopped and the line automatically connected to an evacuated vessel which provided the necessary force for withdrawing liquid. The advantage of this system was the minimal dead volume and continuous cleaning of the sampling line when not in use.

The denitrated solution was usually cooled overnight. If only nitric acid had been employed, this was neutralised and discharged to waste. If not, the supernatant was filtered and returned to the HAW feed vessel. As expected, the final supernatant volume was always slightly greater than the sum of the aqueous solutions added. The oxalate precipitate in the denitrator was washed with a small quantity of dilute $HNO_3/(COOH)_2$, the wash solution being added to the supernatant. The remaining precipitate was then dissolved and eventually destroyed by boiling with nitric acid and hydrogen peroxide and allowing the distillate to go to waste. Finally, this solution was transferred to the oxalate concentrator where it underwent a further volume reduction. Samples were taken of all the final solutions and their volumes noted.

4.0 EXPERIMENTAL SCOPE

The scope of the experimental programme was to establish design and operating conditions which could be successfully applied in the PETRA facility. Thus, instead of studying the chemical nature of the denitration/oxalate precipitation in great detail, attention was primarily focused on the difficulties created by transferring the process from a laboratory to pilot plant scale and then to a hot-cell environment.

5.0 RESULTS AND DISCUSSION

5.1 Reaction start-up

Great attention has been given to the induction period of formic/nitric acid reactions and the possible hazardous consequences (3, 4). It is known that low temperature or reactant concentrations can lead to increased induction periods. Nevertheless, rapidly increasing the initial nitric acid concentration by utilising a higher dosing rate may not be without its problems. This can result in the instantaneous destruction of considerable reactant quantities accompanied by a large gaseous output and an unacceptably high pressure pulse; obviously, the latter should be avoided in active conditions. The overall aim would appear to be the initiation of the process quickly, if not immediately, with one of the reactants present in only a small quantity. This can be easily achieved.

From a process control viewpoint for highly exothermic reactions, the control of the mixture at boiling presents less problems. In the present case, such a high temperature signifies a sufficient reaction rate and, more importantly, a low induction period.

The problem of reactant concentrations can simply be overcome by feeding the HAW through a single tube situated at the bottom of the denitrator. This results in a local region, rich in nitric acid as compared to the rest of the vessel. A dangerous overall nitric acid build-up is avoided, since the concentrations of both reactants are considerable in a particular zone. This would not be the case if the HAW was well distributed by mechanical agitation. Any mixing due to the boiling action or evolution of gases still leaves the vessel contents essentially heterogeneous.

The above principle can be further improved by utilising a reasonable HAW feedrate which helps to sustain locally a sufficient nitric acid concentration. In fact, too low a flowrate can decrease operational safety especially if noble metals, accompanied by their catalytic activity, are present. The increased kinetics could lead to a rapid depletion of one of the reactants as it would an immediate homogenisation of the two components by an highly efficient mixing procedure. In such a case the reaction may oscillate by stopping only to start again, perhaps violently, when both reactant concentrations are sufficiently high. However, too high an initial flowrate is inadvisable. The inherent pressure surge as the reaction departs would be too great. Also, a high initial flowrate could lead to difficulties if the reaction failed to start after a given time interval since the quantity of HAW or more precisely the amount of nitric acid present in the denitrator would be correspondingly more than with a low starting flowrate.

Overall, a satisfactory solution was obtained by starting with a flowrate of $5dm^3/h$ which was progressivly increased up to the nominal flowrate in about 15 minutes. Under these conditions, the nitric acid during the initial phase is almost immediately consumed, i.e. the rate of destruction and hence, the flowrate of off-gas, is limited by the rate of HAW addition. Thus the operator can control the reaction by ensuring that it always proceeds at a high velocity and by limiting the feedstock.

During this phase of the reaction, two parameters are of interest to the operator: pressure and off-gas flowrate. They can be considered interrelated and serve as an indication that the denitration process has successfully commenced and that it is continuing to proceed.

In all experiments the pressure peaked immediately on HAW addition and then fell to a value slightly above ambient only to rise again as the HAW flowrate was increased. Eventually the pressure reached a maximum, where it remained for some considerable time before subsiding as the reaction rate decreased. A similar response was observed when oxalic acid and salts were present. However, in such cases, the pressure oscillated much more and considerable amounts of foam, albeit unstable, were produced.

The main problem with the off-gas flowrate was its reliable determination without using a measurement system which creates a large pressure drop; such an occurrence would be unacceptable as this could imply a pressure build-up within the denitrator. Initially, the off-gas was measured using a mass flowmeter. The instrument incorporated a small capillary which acted as the flow sensor. A small fraction of the off-gas dissipated energy from the heated capillary thereby giving an indication of the mass flow. Unfortunately, towards the end of the first trial, the flowmeter gave abnormally high readings suggesting the presence of liquid in the capillary. Upon inspection it was found to be blocked with a white solid. This reoccurred in the following experiment and it was concluded that such an instrument was unsuitable .

A possible substitute is a turbine flowmeter. This device worked although it could not detect low gas flows. The reliance on bearings is also a disadvantage especially with dirty gases. As hot-cell equipment should be inherently reliable, a third type of instrument was studied.

A hot-wire anemometer usually consists of a heated wire filament maintained at a constant temperature. In much the same way as the mass flowmeter, fluid movement is sensed by the dissipation of energy. However, the chosen anemometer offers several advantages. Its small cylindrical shape can be composed of exotic materials and provides little area for material deposition. It introduces a negligible pressure drop and can be easily replaced. Although little experience has so far been gained with an anemometer during denitration runs, the few results have shown that such an instrument is sensitive and responds quite rapidly to even small off-gas flowrates. However, the long-term reliability cannot be ascertained at present.

Detection of off-gas formation from the apparent density was also considered. However, density measurements using a pair of dip-tubes purged with argon did not provide reliable or reproducible results. Sometimes, the apparent density hardly changed at the start of the experiment while, on other occasions, the signal oscillated widely. This was probably due to the heterogeneous nature of bubble formation on the glass surface of the denitrator which may not prove such a problem with a metallic reactor.

Once the denitration process had satisfactorily commenced, no difficulties were encountered in completing the HAW addition. Furthermore, HAW feeding could be interrupted and restarted without any loss of safety provided that the HAW was always added to a very hot or boiling denitrated liquid.

5.2 General operation

Formic and nitric acids always react to give a large number of products (5, 6). This is a result of the different reactions which can be expected to proceed simultaneously, if not, proportionally.

$$2HNO_3 + 5HCOOH \rightarrow 5CO_2 + 6H_2O + N_2 \tag{1}$$

$$2HNO_3 + 4HCOOH \rightarrow 4CO_2 + 5H_2O + N_2O \tag{2}$$

$$2HNO_3 + 3HCOOH \rightarrow 3CO_2 + 4H_2O + 2NO \tag{3}$$

$$2HNO_3 + 2HCOOH \rightarrow 2CO_2 + 3H_2O + NO + NO_2 \tag{4}$$

$$2HNO_3 + HCOOH \rightarrow CO_2 + 2H_2O + 2NO_2 \tag{5}$$

If the oxalate precipitation of mainly actinides and rare earths is to be successfully attained, it is necessary to reduce the nitric acid concentration to approximately 0.2M. This in itself is not difficult to achieve. However, a second proviso is that the remaining quantity of formic acid is minimal; a high formic acid concentration could cause problems downstream of the denitrator.

The reaction order of formic-nitric acid reactions is high, not necessarily reflecting the stoichiometry of the overall process. This signifies that the kinetics decrease markedly with lowering concentrations until there is no appreciable rate of acid destruction in dilute solutions. A series of experiments was therefore undertaken to determine the optimum overall molar ratio (OMR) of the two acids, formic/nitric, at the beginning of the experiment, i.e. that ratio which reduces the nitric acid concentration to a sufficiently low value while leaving a minimal amount of formic acid in the supernatant.

Previous studies on small quantities have suggested an OMR of less than 2.0 (6). Initial experiments utilising only the two acids confirmed this as shown in Figure 3 which indicates the acid concentrations after 270 minutes of total reflux. As can be seen, for the simple system, an OMR of about 2.0 appears satisfactory. Interestingly, the number of moles HCOOH required to destroy each mole HNO3 or the reaction ratio (RR) was practically independent of the OMR, varying in the range 1.96-2.03 with a mean of 1.98.

Oxalic acid is required to complex the easily hydrolyzable elements, particularly Zr, while precipitating the insoluble oxalates. However, an excess of oxalic acid, up to five times the stoichiometric quantity necessary to complex or precipitate the cations, has been advised (1). Several denitration experiments were performed with either half or the total recommended value of oxalic acid in the absence of salts, Figure 3. Apparent values of RR were 1.91 and 1.83 for either half or the full quantities of oxalic acid respectively. The concentration of oxalic acid at the end of these experiments was insignificant. This is surprising, since dilute solutions of oxalic/nitric acid do not essentially react. However, this may not be the case when intermediate products originating from formic/nitric acid reactions are present.

Much has been written about the influence of nitrous species in denitration experiments (7, 8, 9). Such ions may also play a role in nitric/oxalic acid systems. A series of experiments was performed by adding dilute solutions of these two acids to sodium nitrite and then boiling under total reflux for several hours. No definite conclusions can be made concerning the overall reaction mechanism between nitric, oxalic and nitrous acids. However, it was shown that pure oxalic/nitric acid systems do not react. In contrast, similar mixtures containing nitrite ions led to a marked decrease in the final quantity of oxalate ions and an increase in the amount of nitrate ions. Furthermore, the remaining concentration of nitrite ions was negligible. Interestingly, if it is assumed that each mole of nitric acid is sufficient to destroy two moles of oxalic acid either directly or through an intermediate species, then the formic/nitric acid reaction ratio falls in the range 1.97-2.01.

In contrast to the above observations relating to oxalic acid destruction a preliminary run using simulated HAW containing only the commoner elements had proven successful; a good precipitate was obtained with only about 10 per cent of the initial oxalic acid destroyed. Obviously, some cations offered protection to the acid. A number of trials narrowed the possibility to iron.

An investigation to determine the quantity of iron necessary to afford protection to oxalic acid was made, Figure 4. Little iron was required. Unfortunately, iron can only be considered as a corrosion product evolving from operations upstream of HAW processing. If special alloys or more exotic metals are used in future reprocessing plants, any such iron may be lacking. If this would be the case, the obvious answer would be to add some suitable reagent to the inactive organic acids prior to HAW addition. Iron is one possibility although this may be construed as increasing the salt content. Another solution was thought to be the use of a nitrite scavenger such as hydrazine (10, 11). Two experiments were performed based on this concept. In both, hydrazine was added to the organic acids prior to feeding any nitric acid. The molar quantity of hydrazine was ten times that of iron previously employed; this was to allow for the expected destruction of the scavenger during the process.

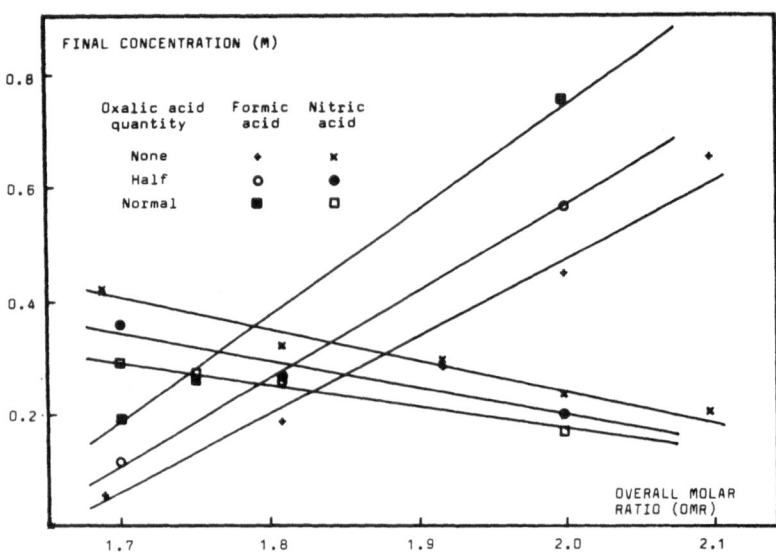

FIGURE 3. THE INFLUENCE OF OXALIC ACID AND THE OVERALL MOLAR RATIO ON FINAL ACID CONCENTRATIONS

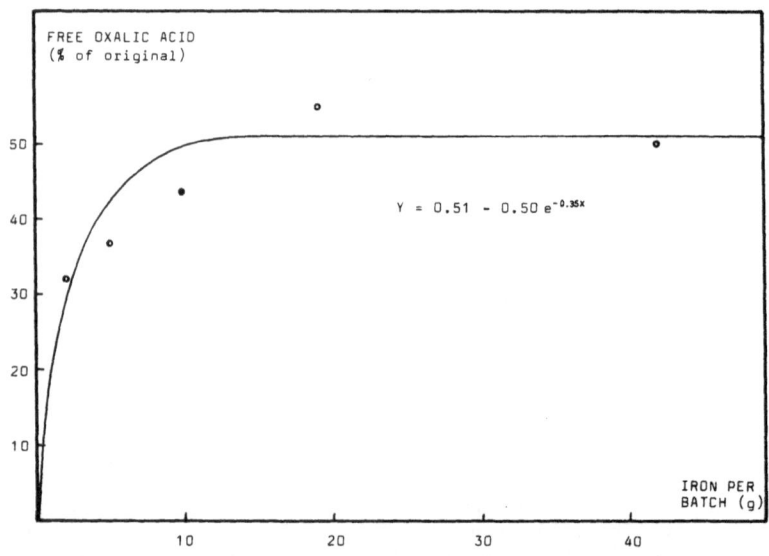

FIGURE 4. THE PROTECTION OF OXALIC ACID WITH IRON

The usual operating procedure was followed during the first run. However, when the denitration reaction failed to start after 2 minutes, feeding of the nitric acid was suspended. A small amount of gas was produced, but in quantities incomparable to that of past experiments. Heating was continued, but no change was observed until after 50 minutes. Then, the contents of the denitration vessel reacted so violently that, within a few seconds, foam completely filled the vessel and condenser and carried over towards the wash column. Fortunately, the foam coalesced just as rapidly leaving the denitration vessel contents docile; later analyses of these contents showed that all of the nitric acid had been destroyed.

The second experiment adopted a similar pattern with only one significant difference. Instead of an induction period of tens of minutes, the reaction commenced after only 5 minutes. This was surprising, since the quantity of reagents employed was comparable. No further tests were made because of the inherent danger of vessel rupture. The second experiment had been accompanied by a pressure pulse exceeding 250mbar compared to the usually 40mbar generally observed, Figure 5.

Several conclusions can be drawn from these observations. Little doubt exists concerning the intermediate role of nitrous acid during the denitration reaction with formic acid. However, the complete prevention of the reaction in the presence of a specific nitrous acid scavenger, such as hydrazine, proves the decisive role of this acid as a catalyst of the reaction. As a consequence, it is indispensible to check the presence of such scavengers in solutions envisaged to be denitrated with formic acid or similar organic reagents to avoid dangerous accumulations of the two primary reactants. Oxalic acid is preferably present in the denitrator from the beginning of the operation in order to allow a homogeneous precipitation of the rare earths/actinides with less impurities. It is adequately protected against degradation by nitrous acid, even in the presence of excess formic acid, only by the contemporaneous presence of iron. It is probable that the protective action of this element is linked to its capability to easily pass from the bi-valent to tri-valent state and vice-versa. Indeed, the other cations which form much stronger complexes with oxalic acid but which exist in only one valency state, completely failed to protect the acid.

A possible way of explaining the role of iron is to assume that it enters in the reaction pathway between nitric and formic acids. The fact that oxalic acid is protected, suggests that iron interacts at the stage of nitrous acid. A possible mechanism may take the form:

$$H^+ + HNO_2 + Fe^{2+} -> Fe^{3+} + NO + H_2O \qquad (6)$$

The tri-valent iron is then regenerated by the formic acid according to the equation:

$$2Fe^{3+} + HCOOH -> 2Fe^{2+} + CO_2 + 2H^+ \qquad (7)$$

Overall, this implies:

$$2HNO_2 + HCOOH -> 2NO + CO_2 + 2H_2O \qquad (8)$$

Unfortunately, the available data are insufficient and too imprecise to appraise this hypothesis remembering that the above reactions would probably be swamped by the primary nitric/formic acid destruction process. Nonetheless, it was observed that the presence of iron led to an increased quantity of NO and diminution of NO_2 in the off-gas as would be expected from Eqs. 2 and 8.

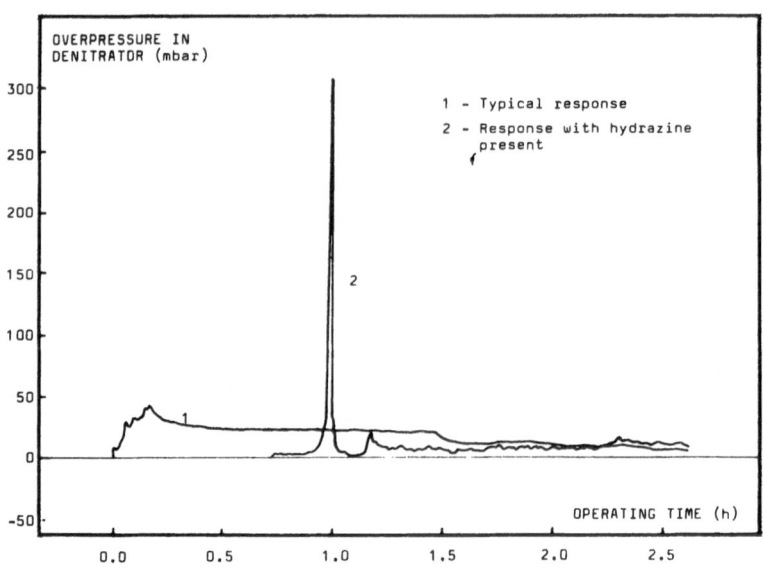

FIGURE 5. THE INFLUENCE OF HYDRAZINE DURING THE INITIAL PERIOD OF DENITRATION

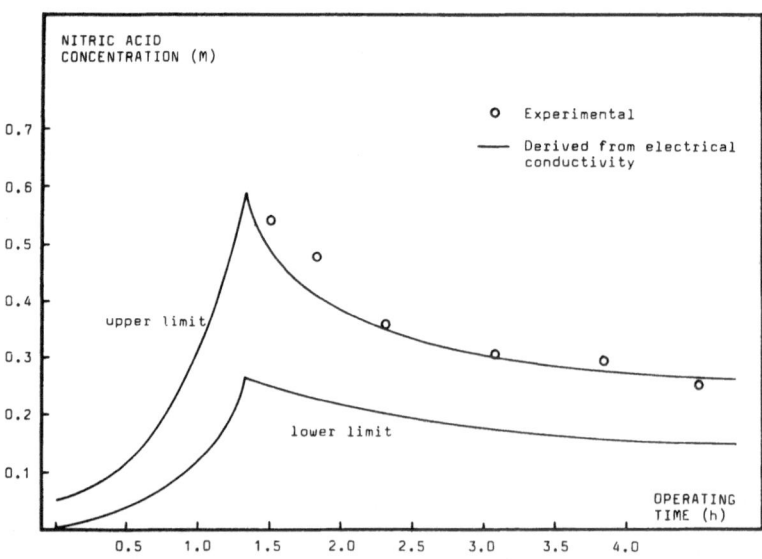

FIGURE 6. THE ESTIMATION OF NITRIC ACID CONCENTRATION FROM ELECTRICAL CONDUCTIVITY

5.3 Reflux period

The initial series of experiments immediately showed that reflux periods of 2h are insufficient. Although the kinetics of nitric acid destruction decrease rapidly with falling reactant concentrations, significant quantities of the acid may still be destroyed under such circumstances. A reflux period of approximately 5h appears to be the most satisfactory. This reflects a total operating time of 7-8h, i.e. a single day-shift, and, with suitable starting conditions, can lead to final low concentrations of both formic and nitric acids.

Throughout the denitration process, the most suitable parameter for following the course of the reactions is the pH. Nevertheless, the reliable measurement of pH in hot-cells without the introduction of foreign ions is not obvious. Instead, attention was focused on a related variable, electrical conductivity. The conductivity of nitric acid solutions is hardly enhanced with the limited presence of salts or weakly dissociated acids such as formic or oxalic. Unfortunately, the heterogeneous nature of the denitrated solution during HAW addition leads to errors. Moreover, any vapour between the electrodes reduces the apparent conductivity drastically. This results in an oscillatory signal. However, the upper limit of these oscillations reflects practically pure liquid and, when compared to the experimental nitric acid concentration, proves sufficiently accurate, Figure 6. Similarly, the nitric acid concentration based on the conductivity immediately at the end of the reflux period was accurate to ± 10 per cent.

An attempt was made to describe the denitrator with a two-compartment model to calculate nitric and formic acid concentrations throughout the period of feeding and then boiling under reflux. The model assumes a small compartment into which the HAW is introduced and a larger compartment representing the bulk of the reagents. The volume of the former compartment is taken as a simple function of the HAW feedrate. Transfer of liquid between the two compartments is also allowed for. Empirical parameters, such as the rate of liquid transfer between compartments and the dependence of the smaller compartment volume on HAW feedrate, were optimised for the present apparatus.

Two equations were used to relate formic and nitric acid concentrations, C_F and C_N, to the rate of nitric acid destruction, r:

$$r = kC_F^2 C_N^3 \tag{9}$$

$$r = kC_F^2 C_N \tag{10}$$

The former was found to fit the experimental data well, except for the later part of the reflux period, where low reagent concentrations are encountered. For this region, a second equation based on the mean stoichiometry of the chemical reaction was found to be more satisfactory. In both cases, the corrected values of the reaction constant, k, were utilised (12).

The influence of oxalic acid, iron and other salts in HAW was also included in the overall model.

Figure 7 shows the expected concentration profile and experimental data for the reflux period of a denitration experiment.

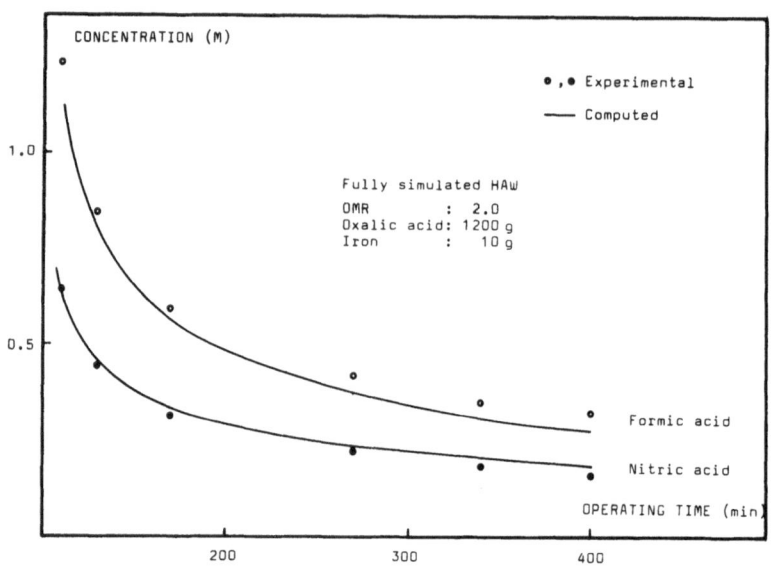

FIGURE 7. A COMPARISON BETWEEN EXPERIMENTAL AND COMPUTED CONCENTRATION PROFILES

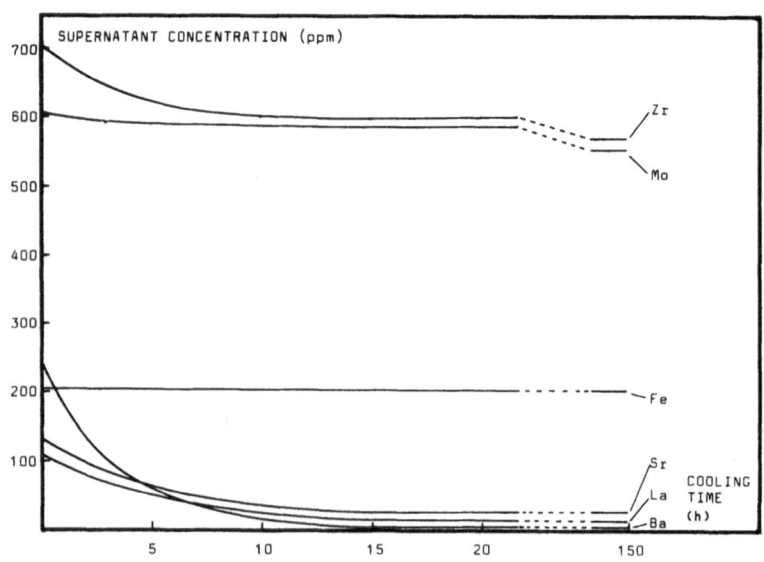

FIGURE 8. THE VARIATION OF SUPERNATANT COMPOSITION DURING THE COOLING PERIOD

108

5.4 Cooling_and_precipitate_recovery

Natural cooling of the supernatant to effect the precipitation process is insufficient. An excessively long period is required and this could lead to the post-precipitation of instable elements such as molybdenum; a small quantity of this element invariably settled out from the supernatant during the first few days even when the solution was maintained at a few degrees Celcius. Secondly, the final temperature may be too high to achieve optimum actinide-fission product separation. It should be remembered that the primary goal of this operation is to produce an actinide-free supernatant. Thus, forced cooling is necessary to ensure that all of the actinides precipitate, even if this results in considerable quantities of impurities such as Ba or Sr contaminating the solid. These elements can be removed in later solvent extraction cycles of the PETRA facility.

Cooling was produced with a coil through which water flowed, placed low inside the denitrator. By this means, final supernatant temperatures in the range 5-10 °C were attained. During the initial cooling period, the oxalate salts which precipitate did so in a reasonably ordered manner. Figure 8 shows how the supernatant concentration of some elements varied with time, while Table I indicates the overall split between the supernatant and precipitate.

Element	Supernatant (per cent)	Precipitate (per cent)
Sr	13.3	86.7
Y	0.3	99.7
Ba	1.8	98.2
Fe	99.0	1.0
Zr	65.0	35.0
Mo	>99.5	<0.5
Ru	100.0	–
Rh	100.0	–
Pd	100.0	–
La	1.9	98.1
Nd	0.2	99.8
Eu	<0.1	>99.9
Gd	<0.1	>99.9

Table I. Distribution of various elements in supernatant and precipitate.

One of the main difficulties throughout the whole process has been the recuperation and destruction of the oxalate precipitate. Unfortunately, the salts not only deposit on the bottom of the reactor, but also on the internal cooling coil, instrumentation, etc. up to a height corresponding to the initial level of the supernatant. The majority of the precipitate is of a hard, crystalline nature, although some white solid appears quite powdery.

The original procedure envisaged utilising a small quantity of boiling, concentrated nitric acid and hydrogen peroxide to destroy the precipitate. Experiments on this basis proved fruitless, since the small amount of returning condensate was unable to attack or redissolve the solid. Later tests, employing larger acid quantities and correspondingly greater condensate flowrates, were equally uneffective. Indeed, the only satisfactory method has been to commence with a nitric acid solution at least equal in volume to the quantity of supernatant removed. On heating, the majority of salts are either dissolved or suspended. The solution is then slowly concentrated by evaporation. As the liquid level falls, the turbulent liquid/air interface helps remove in the vicinity, any solids remaining on vessel internals.

Most of the salts are converted to a nitrate form after many hours of boiling in a regime of moderately concentrated nitric acid. Nevertheless, a significant quantity of fresh precipitate appears, when the final solution is cooled. This material has been separately analysed and found to essentially consist of barium, zirconium and strontium with lesser amounts of other elements including rare earths. Moreover, the solid is not completely attacked by aqua regia, although it is soluble in hydrofluoric acid. Tests are currently being performed to ascertain the form of this solid.

6.0 CONCLUSIONS

The present experimental programme has succeeded in identifying several problems which could arise on transferring the denitration/oxalate precipitation operation from a laboratory to a pilot plant scale within a hot-cell environment. The original design criteria have generally proved satisfactory, while the experience and knowledge gained from the many experimental runs have demonstrated that the overall process can be conducted in a safe and reliable manner. Although some aspects concerning the chemical nature of the process have been clarified, uncertainties still remain in specific areas of the chemistry involved.

7.0 REFERENCES

1. Mannone F., Dworschak H. (Editors), "Chemical separation of actinides from high activity liquid wastes", C.E.C.-J.R.C. Ispra Establishment, S.A./1.07.03.84.02 (1984).
2. Dworschak H., Girardi F., "PETRA, a hot-cell facility for waste management studies", C.E.C.-J.R.C. Ispra Establishment, Second European Conference on Radioactive Waste Management and Disposal, Luxemburg, 22-26 April 1985.
3. Bradley R.E., Goodlett C.B., "Denitration of nitric acid solutions by formic acid", DP-1299 (1972).
4. Holze K. et al., "Aufstellung eines Reaktionsmodells fuer die Denitrierung von Abfallstroemen aus der Kernbrennstoff-Aufbereitung", Chem. Ing. Tech. 51 (7), 754 (1979).
5. Healy T.V., "The reaction of nitric acid with formaldehyde and with formic acid and its application to the removal of nitric acid from mixtures", J. Appl. Chem. 8, 553 (1958)
6. Cécille L., Tanet G., "Etude de l'influence de quelques paramètres sur la réaction de dénitration d'une solution HAW simulée a 5000 l/T", C.E.C.-J.R.C. Establishment, Technical Note No. 1.07.03.80.98 (1980).
7. Kubota M. et al., "Effects on nitrite on denitration of nuclear fuel reprocessing waste with organic reductants", J. Nucl. Sc. Tech. 16 (6), 426 (1979).
8. Orebaugh E.G., "Denitration of Savannah River plant waste streams", DP-1417 (1976).

9. Longstaff J.V.L., Singer K., "The kinetics of oxidation by nitrous acid and nitric acid. Part II. Oxidation of formic acid in aqueous nitric acid", J. Chem. Soc. 1954, 2610 (1984).
10. Biddle P., Miles J.H., "Rate of reaction of nitrous acid with hydrazine and with sulphamic acid", J.Inorg.Nucl.Chem. 30, 1291 (1968).
11. Perrot J.R., Stedman G., "The kinetics of nitrite scavenging by hydrazine and hyrazoic acids at high acidities", J.Inorg.Nucl.Chem. 29, 325 (1977).
12. Study contract, "Studio della neutralizzazione di M.A.W. per mezzo di acido formico condotta in reactore semicontinuo", C.E.C.-J.R.C. Ispra Establishment, No. 2267-83-12-ED-ISP I.

ABATEMENT OF THE NOx EVOLVED DURING DENITRATION OF REPROCESSING CONCENTRATE.

A. DONATO
ENEA CRE-Casaccia   Dipartimento Ciclo del Combustibile – Divisione
MEPIS-RIFIU , Rome, Italy.

Summary

In the treatment and solidification of medium and high level radioactive waste concentrates, large amounts of NOx are produced, which are normally scrubbed with nitric acid or alkaline solutions. In this way big volumes of secondary wastes containing high nitrate concentrations, which cannot be easily discharged, are obtained. The NOx catalytic abatement with selective ammonia reduction, which has been studied at the COMB/MEPIS-RIFIU Laboratory, would permit to avoid this problem, with large cost savings. The NOx catalytic abatement has been experimentally studied at laboratory scale taking into consideration the following operational parameters : the catalyst bed temperature; che gas residence time; the vapour concentration; the NOx concentration; the gas velocity; the catalyst grain size distribution; the catalyst time-life. The best operational conditions have been selected for obtaining abatement yields of the order of 99.5 %.

1. INTRODUCTION

In the reprocessing of the spent nuclear fuel, large amounts of nitrogen oxides are produced. These gaseous oxides, generally named NOx, are normally composed of NO and NO_2, with sometimes small amounts of N_2O. They are produced during the spent fuel dissolution, the oxide conversion of Uranyl nitrate, the medium and high level radioactive waste treatment and conditioning. In the last case the NOx are generated during the waste concentration step, before their solidification, when organic reductants, such as formaldehyde or formic acid, are added in order to lower the nitric acid concentration. Moreover, during the calcination or the vitrification of the concentrated high level wastes, all the nitrates are also decomposed to NOx.

In all cases the produced gas contains up to several percents of NOx, and requires an additional treatment in order to remove the nitric acid and to lower the NOx concentration below the limits established by the laws for the discharge to the atmosphere.

The most used methods for this treatment consists of the gas scrubbing by means of cooled water and/or diluted nitric acid, if the nitric acid recovery is needed, or by means of alkaline solutions. The efficiency of the NOx removal using these methods is, in the best operational conditions and with well designed apparatus, of the order of 90%, being much lower, that is 60≈70 %, in several cases [1][2][3] .

Moreover , the use of the absorption techniques gives rise to considerable amounts of secondary radioactive wastes, containing relatively high concentrations of nitrates, which in turn are to be treated and conditioned if they cannot be discharged as is the case when they do not comply with the requirements of the environmental protection laws against the chemical pollution.

In order to overcome these problems, a study on a selective catalytic abatement of the NOx with ammonia has been performed at the ENEA COMB/MEPIS-RIFIU in the frame of an EEC research contract [4][5]. In this paper the main results obtained in this study are recalled and reported.

2. THE NOx CATALYTIC ABATEMENT PROCESS.

The catalytic chemical abatement of NOx is a process under development at a pre-industrial stage in several laboratories and industries [6][7][8]. The process considered in our study is carried out by means of the chemical reduction of NOx with ammonia according to the following main chemical reactions:

$$6 NO_2 + 8 NH_3 \longrightarrow 7 N_2 + 12 H_2O$$
$$6 NO + 4 NH_3 \longrightarrow 5 N_2 + 6 H_2O$$

over Hydrogen Mordenite extrudates (Zeolon 900 H , a product of the Norton Chemical Co.). The final products are simple Nitrogen and water.

In order to evaluate the process performances at laboratory scale and the influence of the involved phisico-chemical parameters, an experimental loop has been used, reproducing the denitration by formaldeyde of the Eurex medium level waste as NOx source.   The experimental plant was composed indeed of :

- the denitration section, where the simulated waste was continuously denitrated with formaldeyde at the boiling temperature and at the ratio $HNO_3/CH_2O = 2,5$. The molecular ratio [HNO_3 destroyed]/[CH_2O used] has

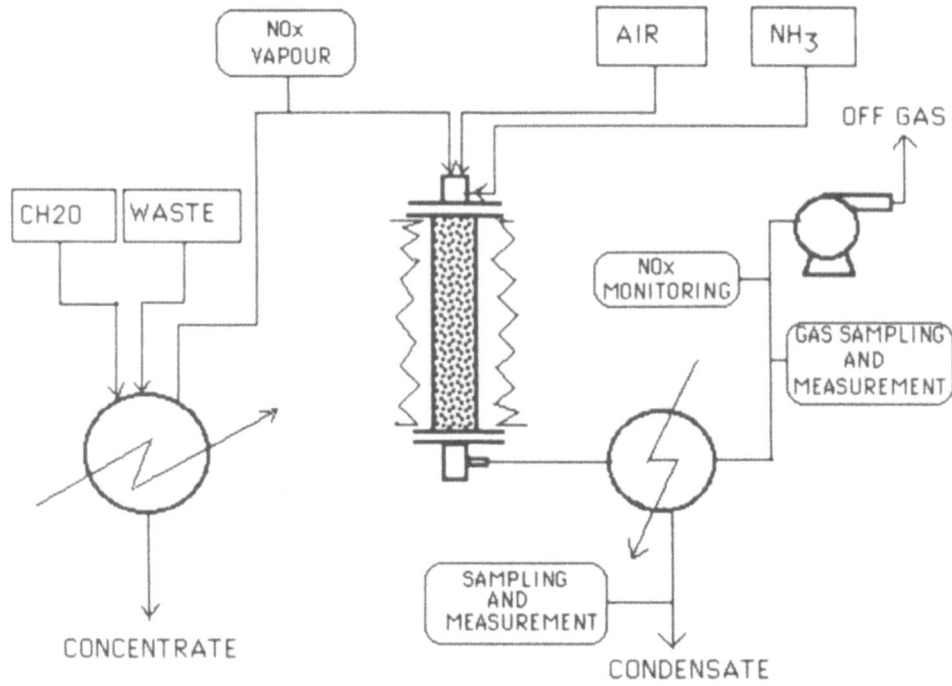

FIGURE 1.   THE SIMPLIFIED SCHEME OF THE NOx CATALYTIC
ABATEMENT PROCESS.

been determined to be constant and equal to 2 during the whole denitration
process.
- the catalytic abatement section , formed by the catalytic reactor and the
ammonia metering and inlet system;
- the analytical section, composed of the sampling and analyzing system,
the cold traps, the vent control.
In the Fig.1 the scheme of the experimental apparatus is shown.
All the experiments   have   been   carried   out   using the molecular
ratio : $NH_3/NOx = 1,0 \pm 0,1$

which seemed to be, on the basis of thermodynamic considerations, the
theoretically most favoured ratio, taking also into account the operational
temperatures.
The catalyst considered in the study has been the Mordenite, which

114

is a zeolite, chemically defined as the Na-SilicoAlluminate $NaSi_5AlO_{12}$. $3H_2O$. The Zeolon 900H, a syntetic hydrogen form of the mordenite commercialized by the Norton Company, has been used.

The NOx catalytic abatement process has been studied taking into consideration the influence on the abatement efficiency of the following parameters :

- The catalytic bed temperature.
- The gas residence time in the catalytic reactor.
- The vapour concentration in the inlet gas.
- The NOx concentration in the inlet gas.
- The NH_3/NOx reaction ratio.
- The gas velocity through the catalytic bed.
- The catalyst grain size.
- The catalyst life-time.

3. THE PROCESS PARAMETERS INFLUENCE ON THE NOx ABATEMENT EFFICIENCY.

3.1 The catalytic bed temperature.

The chemical reduction of the NOx with ammonia is remarkably influenced by the temperature : as the temperature increases, the reaction yield also increases, at least in the range 300÷500 °C. This behaviour, which is in good agreement with the available themodynamic data [9], is clearly shown in the Fig.2 , where the NOx abatement percentage is plotted vs the temperature in the experimental conditions shown in the figure label.

The 99.5 % of NOx is destroyed at 500 °C to form the harmless and not reactant Nitrogen and water, the efficiency being about 99% at 450°C.

When the vapour and the NOx concentrations in the feeding gas are very high, the positive influence of the temperature is remarkably lower. This effect is shown in the Fig.3, where the abatement efficiencies have been determined in the case of a feeding gas having a vapour concentration of 80±5 % V/V and a NOx concentration of 25000 ppm. This composition is nearly the same as the composition of the gas produced in a denitration unit and sent directly to the catalytic reactor without air dilution. In these conditions, the abatement efficiency is less influenced by the temperature and shows a maximum of 93,8% at 450 °C , decreasing then as the temperature increases.

3.2 The gas residence time in the catalytic bed.

The gas residence time is an important parameter both from the kinetic point of view and for a good design of the system . Its influence

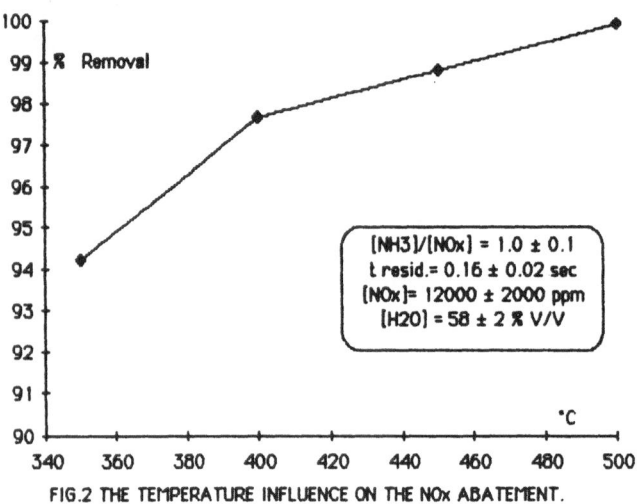

FIG.2 THE TEMPERATURE INFLUENCE ON THE NOx ABATEMENT.

The inset in Fig.2 reads:

[NH3]/[NOx] = 1.0 ± 0.1
t resid.= 0.16 ± 0.02 sec
[NOx]= 12000 ± 2000 ppm
[H2O] = 58 ± 2 % V/V

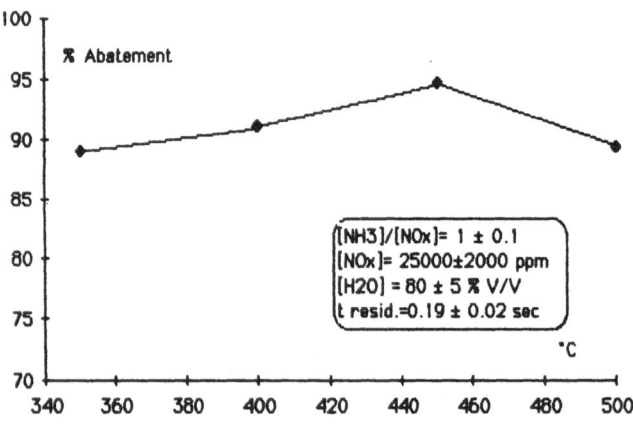

FIG.3 THE TEMPERATURE INFLUENCE ON THE NOx ABATEMENT UNDER
EXTREME EXPERIMENTAL CONDITIONS.

The inset in Fig.3 reads:

[NH3]/[NOx]= 1 ± 0.1
[NOx]= 25000±2000 ppm
[H2O] = 80 ± 5 % V/V
t resid.=0.19 ± 0.02 sec

on the NOx abatement process has been evaluated using a feeding gas containing 20.000 ppm of vapour, and maintaining all other physico-chemical parameters the same. The results obtained in the experimental work are shown in the Fig.4, where the abatement efficiency is plottet Vs the gas residence time in seconds.

In order to obtain removal efficiencies higher than 90 %, a minimum residence time of about 0.02÷0.03 seconds is required. Below this value

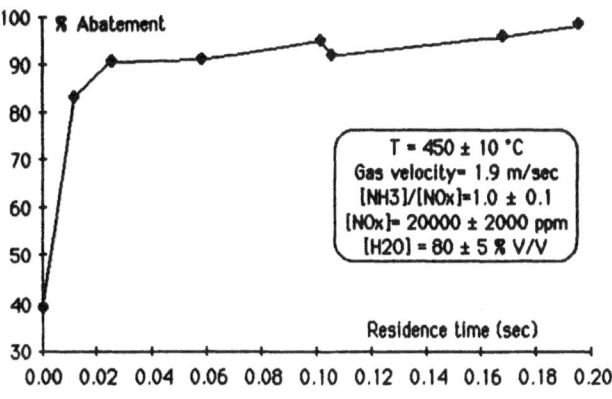

FIG.4 THE GAS RESIDENCE TIME INFLUENCE ON THE NOx ABATEMENT.

the efficiency rapidly drops down, being neverthless 40% when the catalytic reactor is empty.

At residence times longer than 0.02+0.03 seconds, the abatement efficiency increases linearly with the time, that is the amount of destroyed NOx increases as the contact time with the catalyst increases.

From the kinetic point of view, the residence time effect can be explained by the existence of two different mechanisms : below 0.02+0.03 seconds the gas molecules can react on the surface of the catalyst only, and the overall process velocity is controlled by the molecules transfer from the gas core to the catalyst surface . In addition , as a second step, when the residence time is higher, the gas molecules have the possibility to enter the catalyst pores, as more as the residence time is longer. The process efficiency in this case can increase up to the maximum values.

3.3 The vapour concentration in the feeding gas.

The vapour concentration in the feeding gas can strongly influence the abatement process for two main reasons: first, the water is the final product of the NOx chemical reduction with ammonia, and its presence in the gas can act in terms of mass action; second, the water molecules, in consideration of their high polarity, could preferentially occupy part of the catalyst reaction sites , making them not available to the NOx and NH3 reacting molecules.

The experiments carried out have demonstrated that these effects do really exist, as shown in the Fig.5. In the experimental conditions shown in the figure label, the NOx removal efficiency decreases from 99.75% at a vapour concentration of 58% V/V, to 94.1% for a vapour concentration of

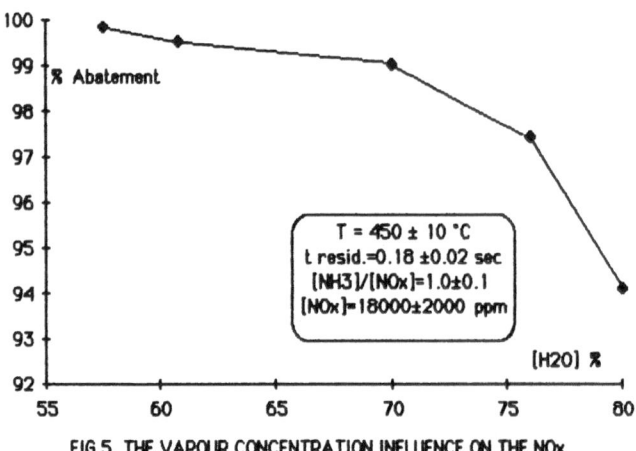

FIG.5 THE VAPOUR CONCENTRATION INFLUENCE ON THE NOx
ABATEMENT.

80% V/V , with a remarkable dropping at about 75% of vapour
concentration. It appears also that, below a vapour concentration of 55%,
the removal efficiency remains at the maximum values.

3.4 The NOx concentration in the feeding gas.

The Mordenite catalyzed NOx chemical reduction with ammonia ,
could be highly influenced by the NOx concentration in the gas, in terms of
the catalyst surface available to the gas molecules.

The importance of this parameter from the point of view of a
global evaluation of the NOx destruction process, has been verified by
means of experiments carried out at increasing NOx concentrations and at
high vapour contents in the feeding gas, all the other experimental
conditions remaining the same. The results obtained in this work are
shown in the Fig.6, where in the label all the experimental conditions are
specified. In the range of NOx concentration of 16000÷38000 ppm, the
process yield noteworthy decreases as the NOx concentration increases.
Nevertheless the yield diminution is continuous and it is not abrupt, as it
would be the case if it was due to the shortage of active catalyst sizes.

This behaviour indeed could be explained by the presence of
secondary reactions in the gas phase, like the following:

$$3 NO_2(g) + H_2O (g) \Longleftrightarrow NO (g) + HNO_3(g)$$

In this way some NO_2 is no more available for reacting with ammonia. This
NO_2 fraction is then found as HNO_3 in the condensate, decreasing in this
way the overall efficiency of the process. The same mechanism could also
partially explain of course the effect of the vapour content.

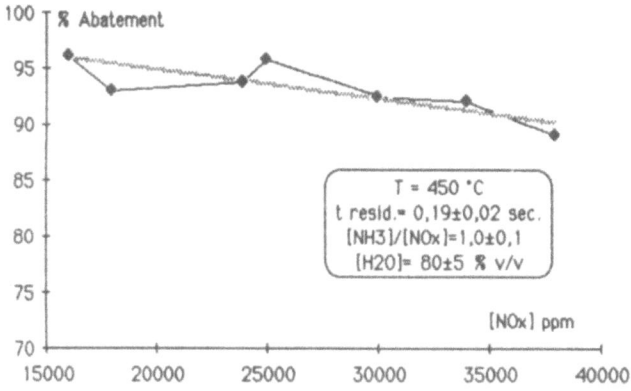

FIG.6 THE INFLUENCE OF THE NOx FEEDING CONCENTRATION ON ITS
CATALYTIC ABATEMENT

3.5 The [NH3]/[NOx] reaction ratio.

The knowledge of the reaction ratio is extremely important in order
to optimize the NOx catalytic destruction avoiding, at the same time, the
release of unreacted ammonia.

The experiments carried out for this purpose have shown that the
real reaction ratio [NH$_3$]/[NOx] is nearly constant and in the range 0.9÷1.05,
being 0.96 on the average. In the Fig.7 the experimentally determined
reaction ratio is plotted vs the NOx feeding concentration : no effects
seem to exist due to the NOx feeding concentration variations.

In all the experiments , which have been performed using a
[NH$_3$]/[NOx] initial ratio of 1.0±0.1, some small amounts of ammonia
therefore have been always found in the condensate. On the average the
95.1 % of the ammonia sent to the catalytic reactor reacted with the NOx.

The reaction ratio seems not to be appreciably influenced by the
other operational parameters, as it is shown in the Fig.8, where the
experimentally determined reaction ratios are plotted against the gas
residence time in the reactor, the vapour concentration in the feeding gas
and the catalyst bed temperature.

3.6 The gas velocity through the catalytic bed.

The influence of this parameter on the process efficiency has been
experimentally demonstrated to be negligeable, at least in the operating
conditions selected for the tests. The results obtained in the experiments
are shown in the Fig.9, where in the label   the operating conditions also
are shown. In the range 0.09÷1.90 m/sec no influence on the NOx removal

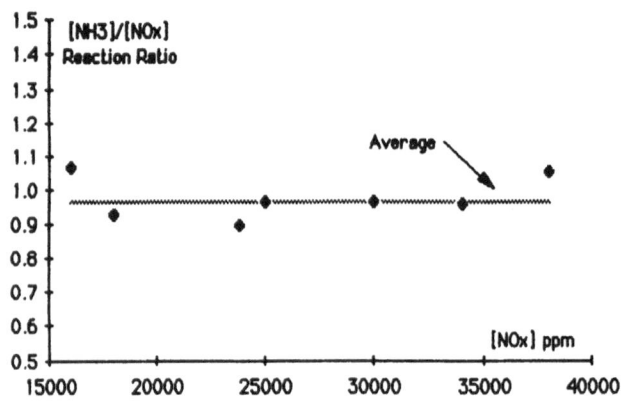

FIG.7 THE MOLECULAR REACTION RATIO VS THE NOx FEEDING CONCENTRATION.

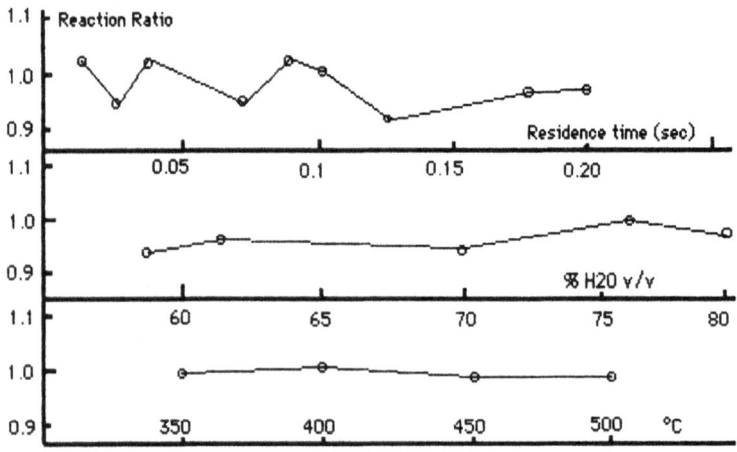

FIG.8 THE INFLUENCE OF PROCESS PARAMETERS ON THE REACTION RATIO

appears.

3.7 The catalyst grain size.

The NOx removal efficiency, as usual in all catalyzed processes, is appreciably influenced from the kinetic point of view by the catalyst surface, and indeed also by the catalyst grain size.

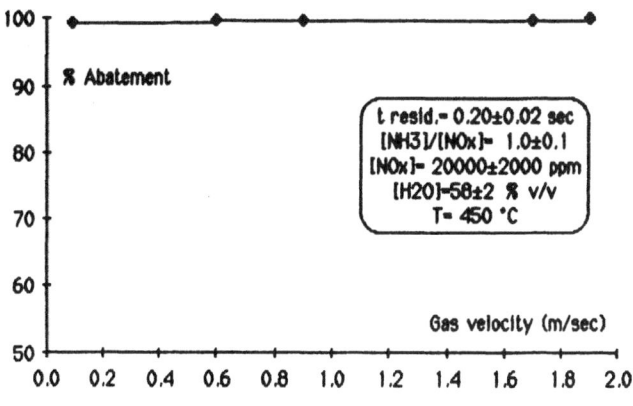

FIG.9 THE GAS VELOCITY INFLUENCE ON THE NOx ABATEMENT.

Three different catalyst grain sizes have been experimentally tested in order to evaluate the effect of this parameter on the process performance: 12÷14 mesh , 18÷20 mesh and 25÷30 mesh, having average grain dimensions of 1.5 mm, 0.9 mm and 0.6 mm respectively. The NOx abatement has been performed in the following operational conditions, which are to be considered as standard for the process: T=450 °C; residence time=0.20±0.02 sec.; vapour concentration=58±2 % v/v; gas velocity=0.09 m/sec.

The NOx removal yield increases as the grain size decreases: it passes from 99.7% at 12÷14 mesh to 99.9% at 25÷30 mesh. Nevertheless the price to be paid for obtaining such an amelioration is a considerable increase of the gas pressure drop across the catalyst bed. The gas pressure drop in fact increases from 10 Hg mm at 12÷14 mesh to 45 Hg mm at 25÷30 mesh.

3.8 The catalyst life-time.

The life-time of the Mordenite catalyst, if it is used in an appropriate way, is relatively long. Its mechanical characteristics and chemical inertness make it very resistant to the action of the corrosive gas passing through, and to the mechanical stresses due to the temperature changes.

The only factor which has been experimentally determined to be very important from this point of view, is the ionic exchange phenomena which could occur when considerable concentrations of salts are present in the vapour and NOx gaseous mixture. In this case in fact the Mordenite crystalline structure could considerably change : the average pore dimension of the Mordenite Hydrogen form is 8÷9 Å while for example, in

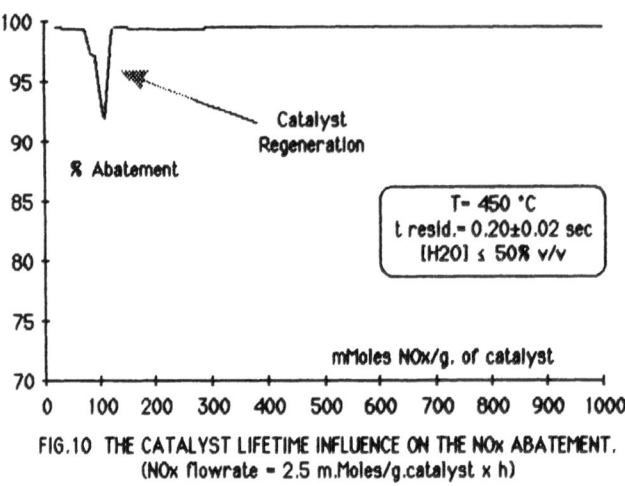

FIG.10 THE CATALYST LIFETIME INFLUENCE ON THE NOx ABATEMENT.
(NOx flowrate = 2.5 m.Moles/g.catalyst x h)

the case of the Sodium form, it is about 7 Å. The catalytic action indeed can be strongly reduced in these conditions.

This effect is clearly shown in the Fig.10, where the experimental performances of the catalyst have been plotted against the test duration up to 400 h of semicontinuous operations, corresponding to a total of about 1.05 moles of destroyed NOx/gram of catalyst. After about 30 h of operations a sudden decrease of the abatement yield has been observed , the efficiency dropping to 92% from the initial 99.5 %. The same catalyst, after chemical regeneration by means of acidic washing , was used again for the duration test, the abatement efficiency being then restored again to the normal value. The 7.2 % of the Mordenite Hydrogen atoms have been demonstrated to have been chemically exchanged with Sodium atoms because of a strong droplets entrainment from the evaporator, producing as described before a change in the Mordenite crystalline structure, and a decrease of the catalyst performances indeed.

4. THE APPLICATION OF THE NOx CATALYTIC ABATEMENT TO THE HLW VITRIFICATION.

As a possible example of an application of the NOx catalytic abatement to a vitrification plant, let us refer to the IVEX Process, a process based on the pot vitrification concept and developed in Italy up to the stage of cold vitrification plant (IVET Plant). The IVEX Plant preliminary design, conceived for the vitrification of the HLW stored at the EUREX Reprocessing Plant, has been commissioned to the French firm SGN, and the studies at present are still in progress.

In the flow sheet of the IVEX Process, referred to the vitrification of HLW coming from the PWR reprocessing, there are two evaporator-denitrators where the NOx gases are produced: the first used for the pretreatment of the stored HLW , the second for the concentration and denitration of the distillates and of the scrubbing towers water.

The other NOx source is the vitrification pot, where all nitrates are decomposed. Four scrubbing towers are foreseen to scrub the gas, the last one acting to lower the NOx concentration to negligible values. The acid produced by the scrubbing is neutralized and , taking into account its low specific activity, it could be considered as low level waste to be discharged. About 800 m^3/y of waste, containing 78 g/l of $NaNO_3$, will be produced in this way.

The EUREX Reprocessing Plant discharges the liquid low level radwastes to the Dora River, in compliance with the radioactivity discharge limits established by the Authorities. The wastes, before discharge, are collected in two ponds of 1000 cu.m each, where they are diluted and controlled. A maximum of 50.000 cu.m about of diluted wastes is discharged to the river each year.

Taking into account the limits established by the Italian law for the nitrate concentration in the discharged effluents [10], that is 121 mg/l as Na nitrate, the   discharge of the secondary wastes coming from the scrubbing would require a dilution factor of 641 times, while the maximum obtainable in the Eurex situation could be 62 about. This means that only the 9,8 % of the scrubbing secondary waste could be diluted and discharged, while the 90,2%, that is 720 cu.m/y, needs to be conditioned in some way.

The situation is completely different when the NOx catalytic abatement replaces the two final scrubbing towers. The condensate from the catalytic reactor, that is 584 cu.m/y, can be discharged at a dilution factor of 53, being possible to dilute it with the existing ponds up to 86 times.

The NOx catalytic abatement allows then to save the additional costs due to the conditioning of the nitrate secondary wastes. As it would not be probably possible to recover the nitrates for not nuclear use (agricoltural use i.e.) for their radioactive contamination, the simplest way to condition them could be a direct solidification by bituminization or cementation. In the case of the bituminization, about 500 drums would be produced each year in this way, at a cost of about 723 $/drum [11], this cost including the production, the trasportation and the burial (costs referred to 1980). A total cost saving of 362.000 $/y therefore, at 1980 prices, would be obtained by replacing the NOx scrubbing system with the catalytic abatement.

5. CONCLUSIONS

The application of the NOx catalytic abatement to the radioactive reprocessing concentrate denitration plants in the place of the scrubbing towers appears to be very attractive both from the environmental protection and from the economic point of view. The process, according to the laboratory experience, seems to work easily on condition that the operational parameters are well controlled. The best operational conditions have been demonstrated to be the following :
- Catalyst bed temperature : 450 ± 10 °C
- Gas residence time : 0,18 ± 0,02 sec.
- Reaction ratio NH_3/NOx = 1,0 ± 0,1

- Gas velocity : 1,5 + 2,0 m/sec.
- Vapour concentration in the gas : ≤ 60 % v/v
An experimental pilot plant is being designed at present, and possibly it will be built to operate in connection with the IVET vitrification plant next year.

REFERENCES

1. PENNAK, H.F. ET AL., " Reduction of Nitrogen Oxides in Vent Gases by Scrubbing." , CONF. 721030 (1977).
2. PULLEY, H., "Fluoride and Nitrogen Oxide Removal from Fuel Cycle Operations", Proc. ANS-AICHE Top. Meeting, Sun Valley, Idaho(1976) KY-L-831.
3. ROMUALD MICHALEK, "Removal of Nitrogen Oxides from Industrial Non-Combustion Sources.", Pollution Eng., March 1976.
4. EEC Research Contract N° WAS-174-81-31-I(S)
5. DONATO, A. and RICCI, G. "Denitrazione di rifiuti radioattivi liquidi a media attività con distruzione catalitica degli ossidi di azoto". EUR 9594 IT (1984).
6. THOMAS, T.R. and MUNGER, D.H. "An Evaluation of NOx Abatement by NH_3 over Hydrogen Mordenite for Nuclear Fuel Reprocessing Plant". ICP-1133 (1978).
7. BRUGGEMAN, A., ET AL. " Elimination of NOx by Selective Reduction with NH_3". Proc. 15th DOE Air Cleaning Conference, (1979).

8. News Features Section, " NOx Controls: Many New Systems Undergo Trials ". Chemical Eng., March 9 (1981).

9. FALCK, F., "Stoichiometry and Kinetiks of the Gas Phase Reaction of Nitrogen Dioxide and Ammonia". Ph. D. Thesis, Princeton University, Priceton, N.J. (1955).

10. Norme per la tutela delle acque dall'inquinamento. Legge Italiana N° 319 del 10 Maggio 1976 (Italian Law).

11. DONATO, A. ET AL., "Radwaste Solidification Modifications by means of Nitrogen Oxides Catalytic Abatement". Proc. Int. Symp. Waste Management 84 , Tucson (USA), March 1984.

DENITRATION OF HLLW FOR ACTINIDE PARTITIONING

L. CECILLE, Commission of the European Communities
(CEC) Brussels, Belgium

M. LECOMTE, Commissariat à l'Energie Atomique
(CEA) Fontenay-aux-Roses, France

Summary

Actinide partitioning from high level liquid waste has been subject
to experimental investigations at the JRC-Ispra and CEA-Fontenay-
aux-Roses. In this context, two main flow-sheets based on solvent
extraction were drawn up which both involved a denitration step by
means of formic acid. In the first process named HDEHP1, denitration
was intended to lower the HLLW acidity down to pH 2 while keeping
actinides and mainly plutonium in a soluble and extractible form
which was proved to be quite effective through secondary reactions
between noble metals and formic acid. As for the second process –
the TBP2 process – denitration was performed during and after the
concentration of high level waste in order to reach a final acidity
of about 1 M HNO_3. On the whole, the results of denitration experiments carried out on real waste at the technical scale agreed quite
well with those previously achieved on simulates in demonstrating
that plutonium can be quantitatively removed from denitration
precipitate under certain operating conditions.

1. INTRODUCTION

In the framework of the research programme of the Commission of the
European Communities on radioactive waste management (1973-1979), the
Joint Research Centre has investigated the possibility of actinide
partitioning from high level liquid waste (HLLW) for transmutation
purposes. In such a context, various flow-sheets based on solvent
extraction techniques have been drawn up and tested on the laboratory
scale, at the JRC-Ispra and then verified on a larger fully active scale
at Fontenay-aux-Roses through a collaboration contract between JRC and
CEA.

The common feature of the two basic flow-sheets considered worthwhile for verification at the technical scale (the HDEHP 1 and the TBP 2
processes) is that they both involve a denitration step of the HLLW which
was proved to play an important part in the implementation of the respective processes.

2. AIM OF THE DENITRATION IN THE HDEHP 1 PROCESS

The main steps of the HDEHP1 process are outlined in figure 1.

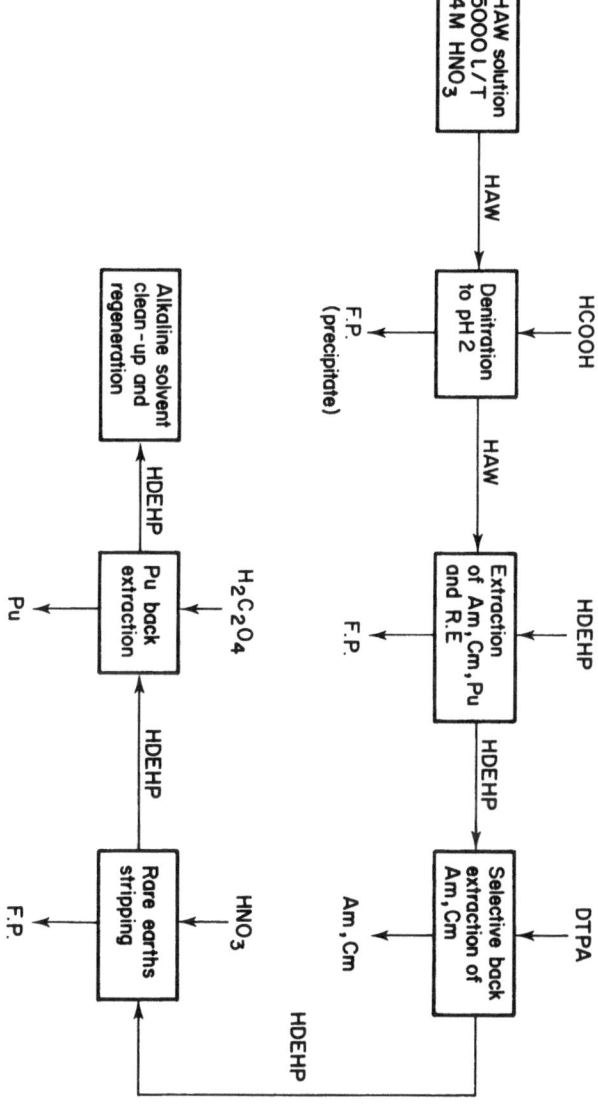

Fig.1 Outline of the HDEHP 1 process

127

In this process, the denitration of HLLW is performed for meeting two different requirements at the same time.

First of all, the denitration step has to reduce the HLLW acidity from 4M HNO_3 down to pH2 to allow the extraction of actinides in 0.3M HDEHP - 0.2M TBP/n-dodecane.

Secondly, the denitration is intended to promote Pu solubilisation (and even depolymerisation) at pH around 2 through reaction between noble metals and the excess of formic acid. This could be particularly helpful if it is assumed that at least part of the Pu contained in the HLLW is not extractible because e.g. already adsorbed on precipitates bound to form during the storage of HLLW[1].

3.   DENITRATION EXPERIMENTS LINKED TO THE HDEHP 1 PROCESS

3.1 Laboratory scale experiments

Basically, the laboratory scale experiments on denitration of simulated HLLW performed batchwise at the JRC-Ispra[2-3-4] consisted of feeding HLLW into an excess of formic acid boiling under reflux (initial $\frac{HCOOH}{HNO_3}$ molar ratio equal to 2) according to the procedure previously set-up by KFK-Karlsruhe[5]. As a consequence, the destruction of nitric acid by formic acid mainly occurred according to reactions (1) and (2).

$$2HNO_3 + 4HCOOH \longrightarrow N_2O + 5H_2O + 4CO_2 \quad (1)$$

$$2HNO_3 + 3HCOOH \longrightarrow 2NO + 4H_2O + 3CO_2 \quad (2)$$

Although no off-gas analysis was performed, the overall stoichiometry of the reaction between HNO_3 and HCOOH was found equal to about 1.7 suggesting that the reactions (1) and (2) occurred in the denitration for 40 and 60% respectively.

The typical variations of pH as a function of the reaction time quoted in figure 2 point out three main steps in the denitration process. The first step which deals with HNO_3 destruction is completed within 2 hours reaction time.

This step was accompanied by a slight change in the colour of the solution and the enhancement of a white precipitate formation. In the second step, due to the reduction of the noble metals to the black state, according to reactions 3 and 4, the colour of the solution turned black and the excess of formic acid (about 0.5M) decomposed according to reaction (5), hence generating hydrogen.

$$HCOOH + Pd^{2+} \longrightarrow 2H^+ + CO_2 + Pd \quad (3)$$

$$3HCOOH + 2Rh^{3+} \longrightarrow 6H^+ + 3CO_2 + 2Rh \quad (4)$$

$$HCOOH \xrightarrow{\text{black Pd/Rh}} H_2 + CO_2 \quad (5)$$

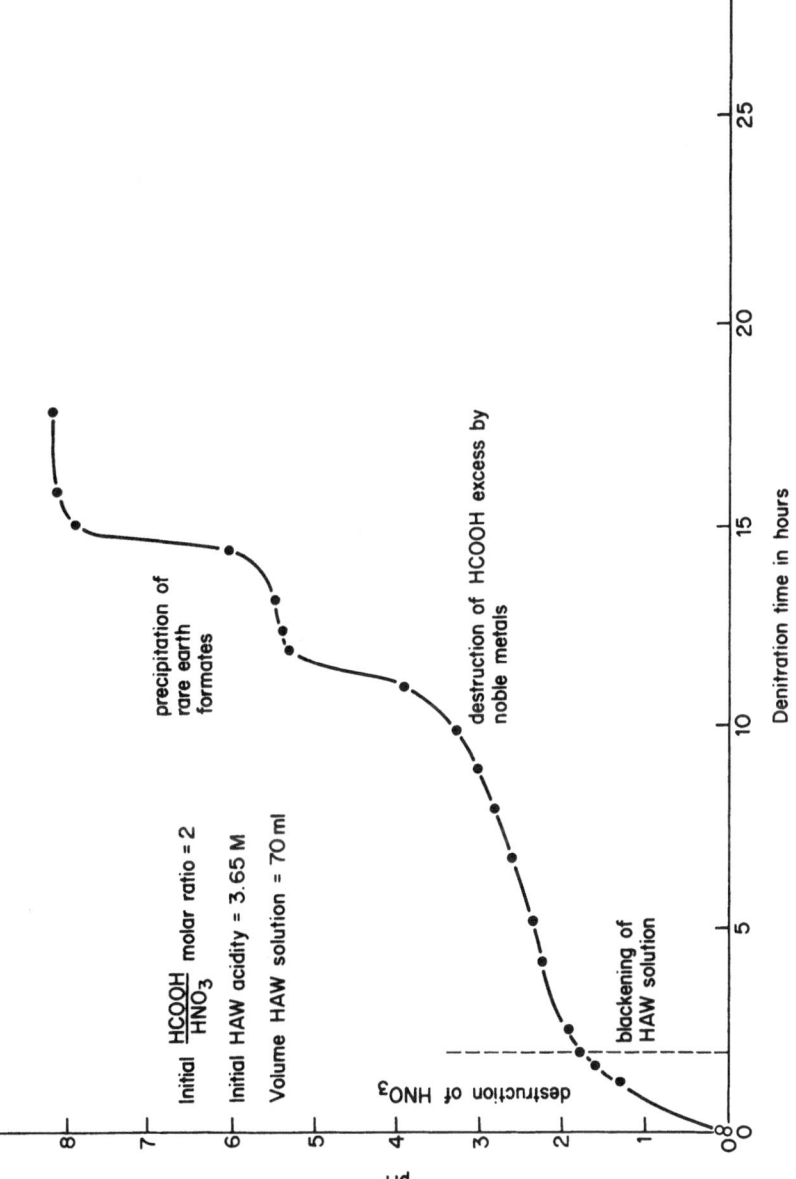

Fig. 2 pH variation of HAW solution as a function of denitration reaction time

The "plateau" at a pH between 5 and 6 might be linked to precipitation of rare earths such as formate salts.

The high final value of pH around 8, observed on this curve, is certainly due to the formation of ammonia, according to reaction (6).

$$2NO + 5H_2 \longrightarrow 2NH_3 + 2H_2O \tag{6}$$

Since as quoted in paper I-2, small variations of pressure may strongly affect the denitration rate, it is worth noting that the related experiments were carried out at a slight underpressure.

Under these operating conditions, the Pu behaviour in solution was found out rather surprising since after a nearly complete and irreversible adsorption on the white precipitate, it progressively returned in solution when the HLLW turned black (fig. 3). Such a phenomenon has been interpreted as the consequence of the hydrogen generation into the solution which reduced Pu (IV) in Pu (III). Similar experiments performed[4] in presence of Pu polymers led to the same experimental observation.

$$Pu\ (IV)\ Polymer + H_2 \longrightarrow Pu\ (III) \tag{7}$$

However, it has to be pointed out that at pH 2, precipitation of readily hydrolysable elements (Sn, Zr, Mo ...) unavoidably occurred which prevented an exhaustive recovery of actinides due to the difficulty to completely wash out the precipitate.

Nevertheless, it was experimentally demonstrated that at least 93% of the Pu could be recovered from the denitrated HLLW at pH 2.5 if a single HCOOH wash of the precipitate is performed.

Since the Pu-solubilisation during denitration has to be linked to the hydrogen generation, which in its turn depends on the concentrations of formic acid and noble metals, it is obvious that a lack of these latter in the HLLW could prevent or decrease the magnitude of such a phenomenon.

3.2. Technical scale experiments

3.2.1 Preparation of the HLLW

The irradiated fuel used for the preparation of fully active HLLW was discharged from the Dutch PWR of Borssele. (Table 1).

The cooling time of the fuel samples at the moment of dissolution was five years and their burn-up was about 33 GW.d.t.$^{-1}$(U). 42 litres of fully active HLLW were obtained by dissolving 11 kg of irradiated fuel and then separating uranium and plutonium by a TBP Purex type extraction cycle.

Before denitration, plutonium, palladium and rhodium nitrates were added to the HLLW to arise their concentrations to respectively; 19, 260 and 80 mg/l in order to adjust plutonium concentration and to ensure the presence of sufficient amounts of noble metals in solution.

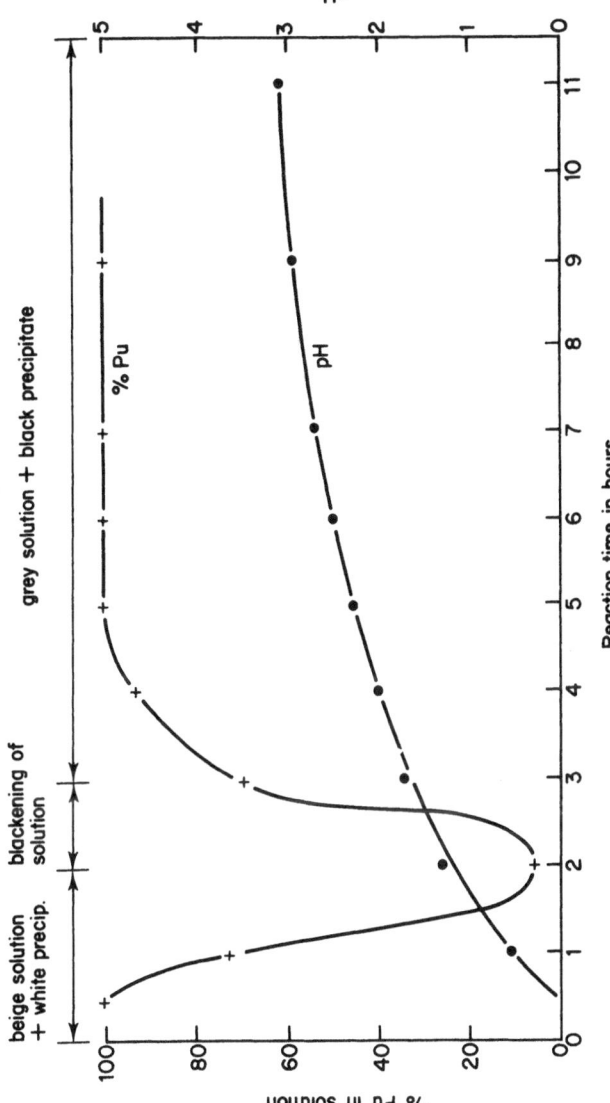

Fig. 3 Denitration of the simulated HAW solution by formic acid variation of pH and % Pu in solution as a function of reaction time $\frac{(HCOOH)}{(HNO_3)} = 2$ initial $(Pu)_{in} = 10$ mg / L HAW volume = 50 mL

Initial HAW acidity = 4.4 M / L

3.2.2 First preliminary denitration experiment

A preliminary denitration experiment was performed by feeding slowly (∼ 250 ml/h) one litre of genuine HLLW into concentrated (26.6 M) HCOOH heated at 80°C using a $HCOOH/HNO_3$ molar ratio equal to 2.

For safety reasons, the denitration was operated under a relative negative pressure of 100 mm H_2O. Surprisingly, the denitration rate was found much slower than expected since after 18 hours boiling under reflux and further addition of formic acid, a pH of only 0.3 was achieved.

It is worth noting that the $HNO_3/HCOOH$ mixture was found less reactive in the denitrator than in sealed sampling tubes since number of these latter exploded during their transfer into the analytical hot-cell.

On the basis of these experimental observations, further investigations were carried out at the JRC-Ispra which highlighted the important part played by small variations of pressure on the denitration rate under the operating conditions chosen (low feeding rate). This was attributed to the existence of gaseous species as intermediate reaction products, the concentration of which in the mixture governs the denitration rate.

Accordingly, it was thought that the slight negative pressure under which the first denitration experiment was performed, was responsible for the slow-down of the denitration rate. Conversely, the fast reaction rate recorded for the aliquots of $HNO_3/HCOOH$ mixture samples was explained by the overpressure created by the gas evolution in the sealed tubes.

3.2.3 Technical scale experiments

Taking into account the foregoing results, the denitration of fully active HLLW was performed by operating under a relative overpressure of 100 mm H_2O and with a slight excess of formic acid (initial molar ratio $HCOOH/HNO_3$ = 2.2), according to the following procedure.

- Heating to 80°C of 14 liters of 26.6 M formic acid.
- HLLW feeding (42 liters, 4.35 M HNO_3) during 10 hours, with a gradual heating increase in order to reach the boiling point of the mixture (feeding rate: 4.2 l/h).
- Boiling under reflux during 3 hours in order to achieve a pH of about 2.

A pH of 2.25 was obtained and an average reaction stoichiometry of 1.65 (moles of HCOOH consumed per mole of HNO_3) was determined. These results, well consistent with those found at the JRC-Ispra, on denitration of simulated HLLW, confirmed the favourable effect of pressure on the denitration rate.

After cooling, the denitrated solution was filtered on 2 "Millipore" filters (5 and 0.5 microns) and a black solid residue of about 100 ml was separated on each filter.

The alpha emitters sorbed on the denitration precipitate amounted to about 6.6% of plutonium and 0.5% of trivalent actinides (americium + curium) initially present in the HLLW. (Table 2).

The solid residue was digested twice with 4.5 liters of 0.5 M HCOOH boiling under reflux. This enabled the removal of about 75% of the fraction of plutonium, sorbed on the denitration precipitate. Consequently, after HCOOH washing, the residual plutonium still carried over in the precipitate represented only 1.6% of the total Pu inventory.

These results showed that washing by formic acid is a necessary step to significantly reduce the residual actinide fraction fixed on the denitration precipitate.

The digestion of the residue with hydrogen fluoride acid led to a further reduction of actinide losses to less than 0.2%.

The main constituents of the denitration precipitate determined by emission spectroscopy were shown to be Pd, Ru, Rh, Zr, Fe and Mo.

4. AIM OF THE DENITRATION IN THE TBP2 PROCESS

The block-diagram schematizing the principal steps of the TBP2 process is reported in figure 4. As for the HDEHP 1 process, the denitration step mainly aims at lowering the HLLW acidity to such an extent that removal of the trivalent actinides by the 30% TBP/dodecane solvent becomes possible. Since the acidity adjustment to 0.1M HNO_3 can be performed by adding aluminium nitrate deficient in nitrate ions to the concentrated HLLW (compulsory step to allow the extraction of the trivalent actinides by TBP), only a partial denitration down to about 1M HNO_3 is required. Therefore, the control of the denitration step is less stringent for the TBP2 process than in case of the HDEHP 1 process.

In addition to the lowering of the HLLW acidity, as in the foregoing process, denitration by formic acid of the concentrated HLLW is expected to promote Pu-solubilisation.

5. DENITRATION EXPERIMENTS LINKED TO THE TBP2 PROCESS

5.1. Laboratory scale experiments

Starting from a 13 fold concentrated HLLW (simulated), prepared by successive evaporation/partial denitration cycles on the diluted HLLW taking care to keep the HLLW acidity always higher than 4-5M HNO_3 [6], the denitration experiments were carried out batchwise by feeding HCOOH into boiling HLLW. By adopting an initial $\frac{HCOOH}{HNO_3}$ molar ratio equal to 1.5 and keeping the reacting mixture boiling under reflux for 5 hours, the final acidity of the concentrated HLLW was lowered down to about 1M HNO_3.

Surprisingly, the stoichiometry of the denitration reaction was found much higher than expected (about 2 instead of 1-1.5) since under these operating conditions, mainly the reactions (6) and (7) should occur.

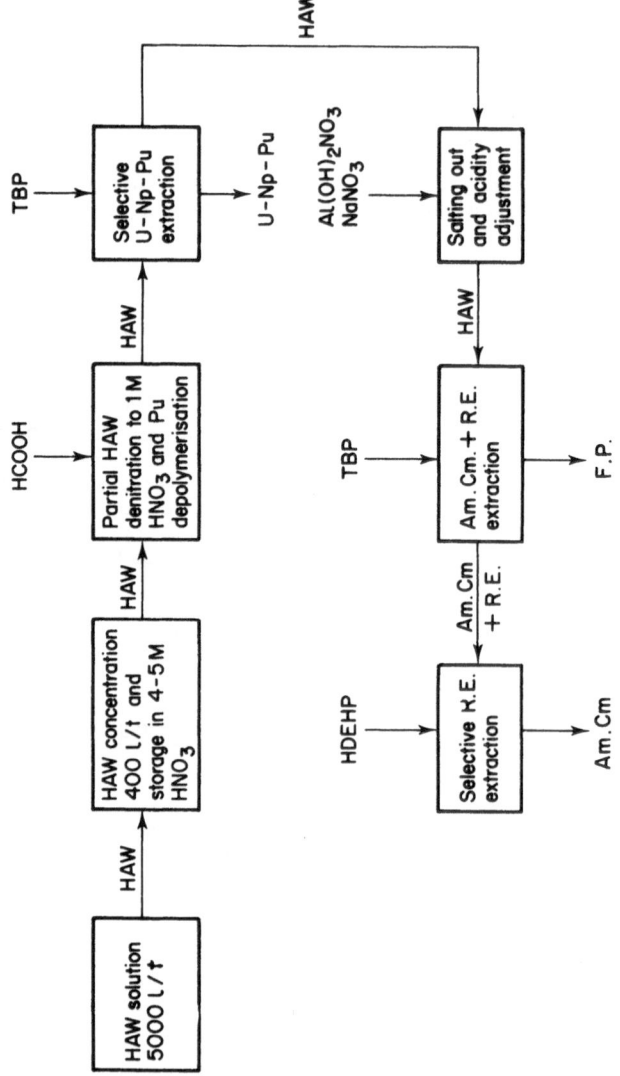

Fig. 4 Outline of the TBP 2 process

$$2HNO_3 + HCOOH \longrightarrow 2NO_2 + CO_2 + 2H_2O \quad (6)$$

$$2HNO_3 + 3HCOOH \longrightarrow 2NO + 3CO_2 + 4H_2O \quad (7)$$

This overconsumption of formic acid could be explained by the action of noble metals, the concentration of which was rather high in the solution (Ru : 5.85 g/l, Rh : 1.04 g/l and Pd : 3.38 g/l). Actually, it was visually observed that during the HCOOH feeding into the boiling HLLW, a slightly black precipitate appeared in the solution for a few seconds every time the stirring was stopped. Moreover the fact that during HCOOH feeding the Pu-solubilisation was showed to be promoted (see later) could indirectly confirm the merits of this hypothesis.

If it is proved that noble metals (in such a range of concentrations) can entail a modification of the stoichiometry of the denitration reaction, the knowledge of their concentration in the HLLW is of paramount importance to choose the right initial $\frac{HCOOH}{HNO_3}$ molar ratio for reaching a predetermined acidity.

When the concentrated HLLW is denitrated to 1 M HNO_3, the whole precipitates (those resulting from the concentration plus those brought by the further acidity lowering down to 1 M HNO_3) represented nearly 7% of the waste volume, hence unavoidably carried over a few percent of actinides. However, it was experimentally demonstrated that by a single wash with 1 M HNO_3, it was possible to remove, on the overall, more than 95% of the Pu and 97% of the Am.

It is worth noting that during denitration, the Pu-solubilisation was promoted especially when Pu is present as polymers into the solution. It is very likely that such a phenomenon results from the action of large concentrations of noble metals on the formic acid since in presence of lower noble metals concentrations, higher and irreversible Pu losses in the precipitates were determined (about 13% instead of 1.5% in the normal case).

5.2. Technical scale experiments

The fully active HLLW used for the verification of the TBP2 process was prepared by the reprocessing of about 10 kg of irradiated Borssele fuel. No noble metal was added for this experiment and 50 liters of 3.3 M HNO_3 HLLW were processed according to the following procedure.

5.2.1 First concentration – denitration step

After a concentration of the HLLW by a factor of 5, a denitration with HCOOH was carried out to lower the nitric acidity from 12.8 M down to 4 M. To this end, 3 litres of concentrated HCOOH (26.6 M) were added (feeding rate: 500 ml/h) to the concentrated HLLW heated to 80°C. The amount of HCOOH added was based on a reaction stoichiometry of 1 (average stoichiometry between reactions (6) and (7).

The solution was then kept at 80°C for 4 hours. Based on the previous experiments, the denitration was operated under a slight overpressure in the reactor vessel, in order to ensure the presence of sufficient amounts of nitrous compounds in the reacting mixture.

A final concentration of 4.7 M HNO_3 was achieved and the acidity balance showed an average reaction stoichiometry of 0.91. This figure agreed rather well with the theoretical stoichiometry of the reaction, but is inconsistent with the stoichiometry determined previously on simulated waste. Probably such a discrepancy has to be attributed to the presence of noble metals into the waste solution in lower amounts than expected.

5.2.2 Second concentration - denitration step

Following a second evaporation step, the concentrated HLLW (4 litres) was again denitrated with HCOOH in order to reach a final acidity of 3 M HNO_3.

For this second denitration, the calculated reaction stoichiometry was 0.98, hence in good agreement with the previous one.

Contrarily to simulated HLLW, the concentration/partial denitration experiments performed on real HLLW did not give rise to any precipitate provided that the acidity is kept higher than 3 M.

5.2.3 Third denitration step

The last denitration step aimed at reducing the acidity of the concentrated HLLW down to 1 M HNO_3. A nitric acid concentration of 1.08 M was obtained and the calculated reaction stoichiometry was about 1.4. After cooling, the denitration solution was filtered and a black solid residue of about 200 ml was shown to carry over about 30% of the Pu.

This residue was digested with 0.5 l of 1 M HNO_3, boiling under reflux for 4 hours. After filtration, the washing solution was added to the denitrated concentrate for actinide recovery by solvent extraction.

In this way, the Pu amount fixed or sorbed on the denitrate precipitate could have been reduced to about 1.8%, demonstrating thus the effectiveness of the washing procedure. Under the same conditions losses of about 0.2% were recorded for americium and curium. These were determined by dissolving a fraction of the washed denitration precipitate in 4 M HNO_3 + 0.15 M Ag^{2+} mixture at room temperature.

6. CONCLUSION

With the operating conditions chosen, it was confirmed that an effective denitration of HCOOH can be achieved if a slight overpressure is applied in the denitrator. This is expected to maintain sufficient amounts of gaseous catalytic species into the reacting mixture.

For the HDEHP1 process for which denitration represents a critical step for actinide recovery, experiments performed on 42 litres of fully active HLLW confirmed the merits of the combined action of noble metals with an excess of formic acid to promote Pu solubilisation at pH 2. Thus, provided that the denitration precipitate is washed out by 0.5 M HCOOH, Pu solubilisation up to at least 98.4 % can be achieved. However, since the noble metals seem to play a key-role in this mechanism, the control of their concentration in the HLLW is of paramount importance.

Concerning the TBP2 process, it must be pointed out that the concentration/partial denitration of 40 litres of real HLLW by a factor of 13, keeping always the acidity beyond 3 M HNO_3 , proceeded according to the basic denitration reactions and did not give rise to any significant precipitate.

However, when denitrating further the concentrated HLLW down to 1 M HNO_3 , precipitates are formed which carried over about 30% of the plutonium. Nevertheless, by simple washings with 1 M HNO_3, about 93.4% of the total Pu could have been removed and subsequently recovered by solvent extraction.

In conclusion, the results of the denitration experiments performed on genuine HLLW at the technical scale (40 l) are in good agreement with those achieved on simulated HLLW at a reduced scale and confirm that, under certain operating conditions, the actinides sorption in the denitration precipitate - which unavoidably appears at low acidity - can be minimized or even reduced to insignificant amounts.

REFERENCES

(1)   P. BARBERO, L. CECILLE, F. MANNONE, G. TANET, S. VALKIERS and
      H. WILLERS.
      "Schémas de séparation des effluents nucléaires de haute activité
      (HAW) faisant appel à l'extraction par solvant. Procédés HDEHP et
      TBP".   Proceedings 2nd Technical meeting on the nuclear transmu-
      tation of actinides. Ispra, Italy, April 21-24, 1980.
      EUR 6929 EN/FR. Session 2.2, p. 211-224

(2)   L. CECILLE, H. DWORSCHAK, B. HUNT, F. MANNONE, F. MOUSTY.
      "Separation of actinides from Purex type HAW raffinates : Develop-
      ment of experimental studies at JRC-Ispra establishment", paper
      presented at the Actinide Separation Symposium. Honolulu, Hawai,
      May 1-6, 1979.

(3)   L. CECILLE, F. MANNONE and F. MOUSTY.
      "The partitioning of high-activity waste (HAW) raffinates".
      Experimental studies on the chemical feasibility of some partition-
      ing processes". Proceedings of IAEA/CEC symposium on the management
      of alpha-contaminated wastes. Vienna, 2-6 June 1980.
      IAEA-SM-246/22 - p. 573-585.

(4)   L. CECILLE's contribution for the final report on the chemical
      separation of actinides from high activity liquid wastes. Chapter 3.
      JRC-Ispra report n° SA/1.07.03.84.02. Edited by F. Mannone and
      H. Dworschak (1984).

(5)   S. DROBNIK
      KFK reports nr. 1830 (1973), 1949 (1974) and 2000 (1974).

(6)   L. CECILLE's contribution for the final report on the chemical
      separation of actinides from high activity liquid wastes. Chapter 5.
      JRC-Ispra report n° SA/1.07.03.84.02. Edited by F. Mannone and
      H. Dworschak (1984).

Table 1 : <u>Characteristics of the HLLW solution used for</u>
<u>HDEHP1 process experiments</u>

Equivalent volume :   3840 l/t H.M.
HNO_3 concentration :   4.35 M
Specific alpha activity :   0.85 Ci/l

<u>Actinide concentration (mg/1)</u>

Pu 239 + 240 :   18.8
Np 237   :   22.5
Am 241   :   75.5
Am 243   : 7.7×10^{-3}
Cm 244   :   6.7

<u>F.P.'s Activity (Ci/1)</u>

Ru 106 :   1.35
Cs 134 :   5.65
Cs 137 :   15.28
Ce 144 :   3.53
Eu 154 :   0.52

Table 2 : % Distribution of actinides in various streams generated by denitration of a fully active HLLW (3840 l/t) and treatments of denitration precipitates

STREAM	VOLUME (1)	ALPHA ACTIVITY specific (mCi/1)	ALPHA ACTIVITY total (Ci)	ACTINIDE DISTRIBUTION (%) Pu 239+240	ACTINIDE DISTRIBUTION (%) Am 241+243	ACTINIDE DISTRIBUTION (%) Cm 242+244
Acidic HLLW 4.35 M HNO$_3$	42	847	35.6	100	100	100
Denitrated HLLW 1.3 M HCOOH pH 2.25	48	718	34.5	93.4	>99	>99
1st PPT wash 0.5 M HCOOH	4.2	28	0.12	4.1	0.33	0.30
2nd PPT wash 0.5 M HCOOH	4.4	4	0.02	0.8	0.05	0.05
3rd PPT wash 8 M HNO$_3$ / 0.1 M F$^-$	2	18	0.04	1.64	0.11	0.09
PPT after HNO$_3$ / F$^-$ washing	0.2			<0.2	<0.2	<0.2

OPERATIONAL EXPERIENCE OF THE DENITRATION OF SIMULATED
HIGHLY ACTIVE LIQUID WASTES DURING VITRIFICATION

M C JERVIS
Research and Development Department, BNFL Sellafield.

Summary

This paper summarises the work carried out in the laboratory and on
the Full Scale Inactive Facility at Sellafield aimed at optimising
the calcination stage of the Windscale Vitrification Plant
process. The effects of the calcination conditions and the
calcination additives, sucrose and lithium nitrate, on denitration
and off-gas behaviour are discussed.

I   Introduction

The Windscale Vitrification Plant (WVP), scheduled for operation in
the late 1980's, will have 2 lines, each capable of reprocessing about
145 m^3 of Highly Active Liquor (HAL) per annum.   It will process HAL
arisings from future Magnox and Oxide fuel reprocessing as well as the
current Magnox backlog.   The evaporation and calcination of the HLLW
will be performed in a rotary kiln and a separate melter will be used to
react the calcined product with a borosilicate glass.   The separation of
the evaporation and calcination stage from the melting stage gives the
opportunity for each stage to be optimised separately.
The primary objective of the calcination process development has
been to optimise the denitration reaction to allow the production of a
calcine with the correct reaction kinetics for the melting stage.   A
secondary objective has been to lower volatility during calcination.   As
long as the off-gas treatment enables the authorised discharge limits to
be comfortably achieved, the priority has remained in producing a
chemically reactive calcine by optimisation of calcination additives and
conditions.   All scrub and condensate liquors from WVP off-gas treatment
will be fed back to the calciner either directly (dust scrubber liquor)
or indirectly after evaporation (condensate, NO_x scrubber liquor etc).
This paper will summarise the work carried out in the laboratory
and on the Full Scale Inactive Facility (FSIF) at Sellafield in order to
meet the calcination optimisation objectives.   This work has included
investigation by Differential Thermal Analysis (DTA) of the reaction
chemistry between magnox HAL simulate and the calcination additives,
lithium nitrate and sucrose.   The effects on denitration of calciner
tube temperature, free acidity of HAL simulate and sucrose input have
been examined.
The feasibility of routeing an ammonium nitrate (NH_4NO_3)
bearing stream with HAL arising from the reprocessing of Oxide fuels has
been studied.   The presence of NH_4NO_3, which undergoes highly
exothermic decomposition, has yielded some interesting results.

141

II   Laboratory Scale Calcination Studies

The base glass developed for WVP is a lithium/sodium borosilicate glass. A proportion of the product glass lithium oxide will be added to the HAL as lithium nitrate prior to feeding to WVP. The interactions of lithium nitrate and the two other main feeds to the WVP calciner, Highly Active Liquor and sucrose, were investigated in a series of experiments using Highly Active Liquor Simulate (HALS). Three sets of experiments were carried out and in each set, the HALS/$LiNO_3$ ratio remained constant whilst the sucrose content was varied. The extent of denitration and the level of ruthenium loss were determined. No attempt was made to replicate the geometrical conditions of calcination. The crucible and contents were placed in a furnace for 13 minutes. Temperatures of 300, 400 and 500°C were studied and the contents were at the process temperature under investigation for a mean time of 5 minutes. The levels of lithium nitrate addition to HALS studied were 0, $^1/2$ and 1 when expressed as fraction of product glass lithium.

The levels of ruthenium retention followed the general trend indicated in Figure 1 which shows the level of ruthenium retention for a HALS/Half $LiNO_3$ mixture versus sucrose level at 400°C. Ruthenium retention decreased as the temperature was increased but the variation of the lithium nitrate content produced no significant changes. Figure 1 also shows that as the level of sucrose addition increases above 20-25g of sucrose per 100g of equivalent waste oxides, the ruthenium retention does not continue to increase with increasing sugar content. This indicated that the correct stoichiometry between HAL and sucrose with regard to minimising ruthenium carry-over may be within the region where this levelling of the graph starts.

Results of nitrate analyses showed a decrease in nitrate retention with temperature as expected. Complementary total nitrogen analyses showed that no intermediary nitrites (NO_2^-) were present. Nitrate retention expressed as a percentage, and with a sucrose content of 18g per 100g equivalent waste oxides, ranged from 23% at 300°C to 4.7% at 500°C with all $LiNO_3$ present, and from 19.6% at 300°C to 0.4% at 500°C with no $LiNO_3$ present. Figure 2 illustrates the effect of different $LiNO_3$ contents on nitrate retention at 300, 400 and 500°C. The graph shows that at 300°C, with no $LiNO_3$ present, the residual nitrate was 506 mg. When all the $LiNO_3$ was present, the residual nitrate was 700 mg. Thus assuming that when $LiNO_3$ was present, a similar amount of nitrate from the HALS remains undecomposed, some lithium nitrate must have decomposed or reacted at 300°C as $LiNO_3$ contributed 621 mg of nitrate when all the $LiNO_3$ was present.

Differential Thermal Analysis was carried out on a number of samples in order to give a better understanding of the reaction mechanisms during denitration. The DTA and NO_2 release traces are reproduced in Figures 3a-3d. Figure 3a shows a broad NO_2 trace centred about 400°C and a number of endotherms between 275°C and 400°C. These are indicative of the thermal denitration of the component nitrates as many have denitration temperatures in this region. No liquids can be placed on the thermal balance and therefore, all mixtures were dried to remove the majority of the free nitric acid and water. The big endotherm observed on all traces at 100°C corresponds to the evaporation of the remaining water.

The traces from a HALS/Full $LiNO_3$ mixture (Figure 3b) show 2 additional endotherms and a modified NO_2 peak. The large endotherm at 200^oC is due to $LiNO_3$ melting whilst the endotherm at $475-500^oC$ is probably due to the denitration of some of the $LiNO_3$. This is confirmed both by a new peak on the NO_2 trace at $475-500^oC$ and the reduction of this peak and the associated endotherm when the $LiNO_3$ content is lowered. The modified NO_2 trace at 300^oC suggests a reaction of lithium nitrate as was envisaged from the crucible experiments.

Figure 3c shows the traces obtained from a HAL/sucrose mixture in which the sucrose concentration was 18g per 100g equivalent waste oxides. The effect of sucrose was to lower the NO_2 peak and the corresponding endotherm to about 340^oC. There is also a greater production of NO_2 between $250-300^oC$. This is indicative of a sucrose/nitrate reaction. The HALS/$LiNO_3$/sucrose traces (Figure 3d) show the same trends as the binary mixtures. When the sucrose level is decreased below 18g per 100g equivalent waste oxides, more NO_2 is evolved around 400^oC. Similarly as the sucrose level is increased above the stated level, more denitration occurs at the lower temperatures and less at 400^oC.

The laboratory calcination and the thermal analysis experiments help to explain the effect of sucrose during the calcination stage. It must be re-emphasised that the principal object of the calcination process development has been to produce a calcine with the correct reaction kinetics for the melter stage. To this end, sucrose was added as a reagent whose thermal decomposition would release large volumes of gas which would lead to the formation of a friable calcine and contribute to the prevention of caking of the tube wall. Sucrose was chosen ahead of other additives which would have a similar effect because it was known that it would also reduce ruthenium volatility. However, the sucrose addition will be optimised with regard to calcine reaction kinetics so long as the ruthenium volatility remains within the region where the WVP off-gas treatment system will keep aerial discharges well within the allocated limits. The crucible experiments demonstrate that ruthenium volatility does decrease as the sucrose input is increased whilst the thermal analysis work indicated that increasing the sugar content reduces the denitration temperature of many component nitrates in the HALS. It must be remembered that as the thermal analysis work had to be carried out on a solid sample, no free nitric acid was present. In the calciner the degree of chemical denitration of the metal nitrates by sucrose will be less than observed in the DTA work because the bulk of the sucrose will react with additional nitrate arising from the presence of free nitric acid. The reduction of the ruthenium volatility is a consequence of the chemical denitration by the sucrose. The oxidation of the sucrose creates a more reducing atmosphere which is thought to inhibit the formation of the volatile RuO_4 but the escort mechanism and stoichiometry have yet to be ascertained.

Lithium nitrate had no discernible impact on ruthenium volatility. However, it appears that some of the lithium nitrate is reacting even in the absence of sucrose at $250-300^oC$ which is about 300^oC below its decomposition temperature. Lithium nitrate is added to provide a uniformly distributed fluxing agent for the melting stage but as some is obviously reacting in the calciner, the possibility of a eutectic with one or more of the other metal nitrates needs to be investigated.

The final part of the thermal analysis work showed the danger of feeding only sucrose and lithium nitrate to the calciner. It was found that such a mixture would be the only hazardous combination of feeds resulting from the failure of part of the WVP feed system. Figure 4 shows the DTA exotherm when a $1/2$ LiNO$_3$/sucrose mixture was heated. An energy release could be heard when a crucible containing such a mixture was heated to 300°C in a muffle furnace. Although conditions were obviously not the same as would be encountered in a calciner, the possibility of an explosive chemical denitration of the lithium nitrate by sucrose is one of the reasons why LiNO$_3$ will be pre-mixed with HAL prior to feeding. This will ensure a mixture of LiNO$_3$ and sucrose will never be fed to the WVP calciner.

III Full Scale Inactive Facility Trials

As part of the calcination process development, a number of trials have been carried out on the Sellafield FSIF in which the melter was detached and the calciner was run separately. This enabled the calcine properties and yields to be determined thus giving a better understanding of this stage of the vitrification process. The reactivity of a particular calcine is influenced by the fraction of product glass lithium added as lithium nitrate to the calciner, its particle size distribution and by the residual nitrate content. The optimization of these properties, whilst maintaining volatilities at levels which will keep WVP discharges comfortably within allocated discharge targets after treatment, is the primary aim of the calcination development work.

The first calcination runs were concerned at determining the effect of different calciner tube wall temperatures on the performance of the calciner and on the quality of the calcine. A Magnox HAL simulate was fed at a rate which would give an equivalent waste oxide input of 3750g hr^{-1}. One half of the product glass Li$_2$O was added as lithium nitrate to the calciner and sucrose was added at a level of 18g per 100g of equivalent waste oxides. As all the furnace zone temperatures were changed by the same amount for each part of the experiment, the same temperature profile was maintained along the tube. The furnace temperatures of the four zones started within the range 760-890°C and were reduced in 20°C increments to the range 640-770°C. It should be noted that the internal temperature of the tube will be some 50-100°C lower.

Figure 5 shows that the tube expansion increases linearly with zone wall temperatures. The calcine yield decreased linearly as the temperature was increased (Figure 6). This experiment demonstrated that for a particular input of calcination additives, the thermal denitration conditions can be specified to yield a calcine with the optimum residual nitrate content.

Operational variations in the HAL and Dust Scrubber Recycle acidities will lead to a range of free acid inputs to the WVP calciner. The objective of the next FSIF calcination run was to observe the effect of varying the free acid input from 30 to 80 moles/hour. This reflects the full range of free acid inputs likely to occur during routine WVP operations. A Magnox HAL simulate was used and the equivalent waste oxide, lithium nitrate and sucrose inputs were identical to the previous calcination trial. The four furnace zone temperatures were set between

700°C and 850°C with the intention of producing a calcine with a 20-30% w/w residual nitrate content. Figure 7 shows the relationship of free acid input to residual nitrate in the calcine. It demonstrates that the residual nitrate of the calcine was unaffected by the free acid input. Similarly, free acid input had no measurable effect on the mean particle size of the calcine or the calciner tube expansion. Therefore it can be concluded that even with the lowest free acid input of 30 moles per hour, there is an excess of acid for the sugar/nitric acid reaction and the calcination stage of the WVP process will be unaffected by the unavoidable variation of free acid input to the calciner.

A correlation can be made between calcine mean particle size and tube expansion from the results of those FSIF calcination runs where the equivalent waste oxide, lithium nitrate and sucrose inputs have remained constant. Figure 8 shows that mean particle size decreases with increasing tube expansion up to a tube expansion of 14.7mm. The decrease in mean particle size coincides with decreasing nitrate content because evolution of gas from thermal denitration assists in the breakdown of the calcine. As the heating to the calciner causes the tube expansion to rise above 14-15mm, baking of the calcine causes an increase in mean particle size and this is obviously undesirable for the melting stage.

The latest FSIF calcination trial has investigated the impact on calcine properties and off-gas behaviour of varying the sucrose input. A Magnox HAL simulate was again used and the equivalent waste oxide and lithium nitrate inputs were maintained at the levels of previous FSIF calcination trials. Sucrose inputs were ranged between 0 and 54g sucrose per 100g equivalent waste oxides. The four zone temperatures were set between 700°C and 850°C. Figure 9 illustrates the relationship between throughput and the residual nitrate in the calcine. It was found that nitrate in the calcine decreased linearly as sucrose throughput increased and it confirms the findings of the DTA work which showed that sucrose causes a significant amount of chemical denitration. There is still obviously an excess of nitrate in the region of the highest sucrose throughput studied as the residual nitrate continues to decrease linearly. The same furnace zone temperatures were maintained throughout the run but the tube expansion increased linearly with increasing sucrose throughput (Figure 10). Further studies will be required to determine whether this is as a result of exothermic chemical denitration leading to less heat being required for evaporation of nitric acid and thermal denitration. However, it is an interesting observation and may be a useful phenomenon if a correlation can be proved. After a 10 hour feed period at the highest sucrose throughput, the nitrate content of the dust scrubber recycle liquor was five times lower than after the corresponding period with zero sucrose input. In the latter case, more HNO_3 was distilled from the calciner and there was some condensation in the dust scrubber. At higher sugar throughputs, less HNO_3 was evaporated because of chemical denitration. Although the NO_2 production was much greater at the high sugar throughput, the dust scrubber has a NO_x D_F of approximately 1. NO_2 is removed in the later part of the off-gas system and the nitrate contents of the condensate and NO_x scrubber liquors did show an increase with increasing sugar input (Figure 11).

Dried solids at 1000°C analyses of the dust scrubber recycle, condensate and NO_x scrubber liquors showed that the dust scrubber was about 99% efficient. Therefore the percentage of feed recovered in the dust scrubber is a reliable indicator of the sucrose on the total carry-over. Figure 12 illustrates the relationship between sucrose input and feed carry-over to the dust scrubber as a percentage of the total feed. The percentages of feed Ru and Fe recovered in the dust scrubber are also plotted. As the Fe carry-over is of the same proportion as the total feed carry-over, it is clear that the majority of carry-over occurs by particulate entrainment in the off-gas stream. There was a notable increase in particulate carry-over when the sucrose input went above 25-30g sucrose per 100g equivalent waste oxides. This corresponds to a step change decrease in mean calcine particle size (Figure 13).

The differences between the percentages of ruthenium and total feed recovered in the dust scrubber were due to ruthenium volatility. It can be seen from these differences that the ruthenium volatility decreased by a factor of approximately 6 between zero sucrose input and a throughput of about 28g sucrose per 100g equivalent waste oxides. It is not possible to observe decreasing ruthenium volatility in the dust scrubber at higher sucrose throughputs because the effect was concealed by higher particulate carry-over. However, the percentage of feed ruthenium recovered in the condensate continued to decrease after particulate carry-over became significant indicating that the volatile ruthenium production continued to fall at higher sucrose throughputs (Figure 14).

The optimum sucrose throughput has not yet been decided. However a sucrose throughput above 30g per 100g equivalent waste oxides ie unlikely to be chosen because the step change in particulate carry-over could lead to blockages of the calciner to dust scrubber off-gas pipe during prolonged periods of running. The ruthenium volatility decreased from about 5.6% of Ru fed at a sucrose throughput of 4.5g per 100g of equivalent waste oxides to about 1.5% at the highest sucrose throughput possible before particulate carry-over became a significant factor. It would be desirable to choose the latter throughput but if a significant advantage with respect to calcine reaction kinetics were to be accrued by choosing a lower sugar throughput, the impact on WVP aerial discharges would be very small because of the relatively small differences in volatilities observed at the various sucrose throughputs and the built-in safety margins in the WVP off-gas system. Discharges would still be well within the allocated discharge limits whatever sucrose throughput is set. Calcine reaction kinetics studies with relation to sucrose level are continuing.

It has been seen that sucrose is a suitable calcination additive for Magnox wastes. Preliminary pilot scale calcination trials with Oxide HALS are of interest. It has been decided to vitrify the Thermal Oxide Reprocessing Plant salt-free medium active concentrate with the Oxide HAL. This will result in NH_4NO_3 being present in the WVP feed. NH_4NO_3 undergoes a highly exothermic decomposition but extensive studies have shown that the NH_4NO_3 concentration proposed in the oxide HAL is well within the safety margins and no pressure excursions have been observed in any pilot scale calcination trials with oxide HALS. During the calcination trials, sucrose was fed at a rate of

20-30g per 100g equivalent waste oxides. It was found that when no NH_4NO_3 was present, the Ru volatility was of the same order as observed in the Magnox calcination studies with the same sucrose input. The presence of NH_4NO_3 at its estimated flowsheet level increased the volatility by a factor of three. This is probably because the addition of NH_4NO_3 increased the NO_3^- content by about 20%, thus reducing the effective sugar content. It should be added that Ru-106 will be present at similar levels during oxide and magnox calcination. The effect of NH_4NO_3 on ruthenium volatility requires further investigation but preliminary pilot scale calcination studies, without optimisation of the sugar, show that sucrose will be a suitable calcination additive during the vitrification of oxide highly active liquid waste.

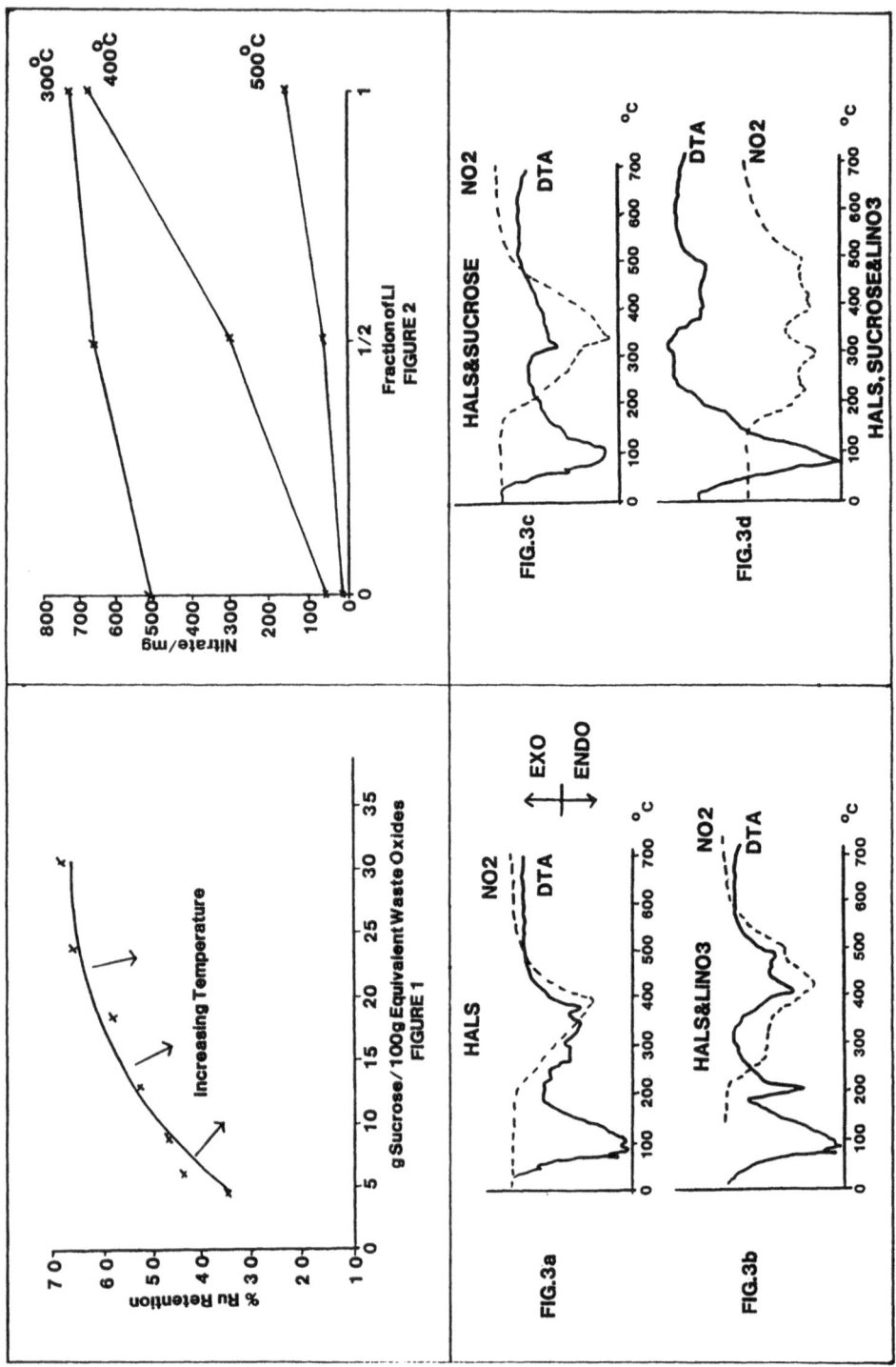

% Ru Retention

g Sucrose/100g Equivalent Waste Oxides
FIGURE 1

Increasing Temperature

Nitrate/mg

Fraction of Li
FIGURE 2

300°C

400°C

500°C

FIG.3a HALS   NO2   DTA

EXO ⟷ ENDO

FIG.3b HALS&LINO3   NO2   DTA

FIG.3c HALS&SUCROSE   NO2   DTA   °C

FIG.3d HALS, SUCROSE&LINO3   DTA   NO2   °C

148

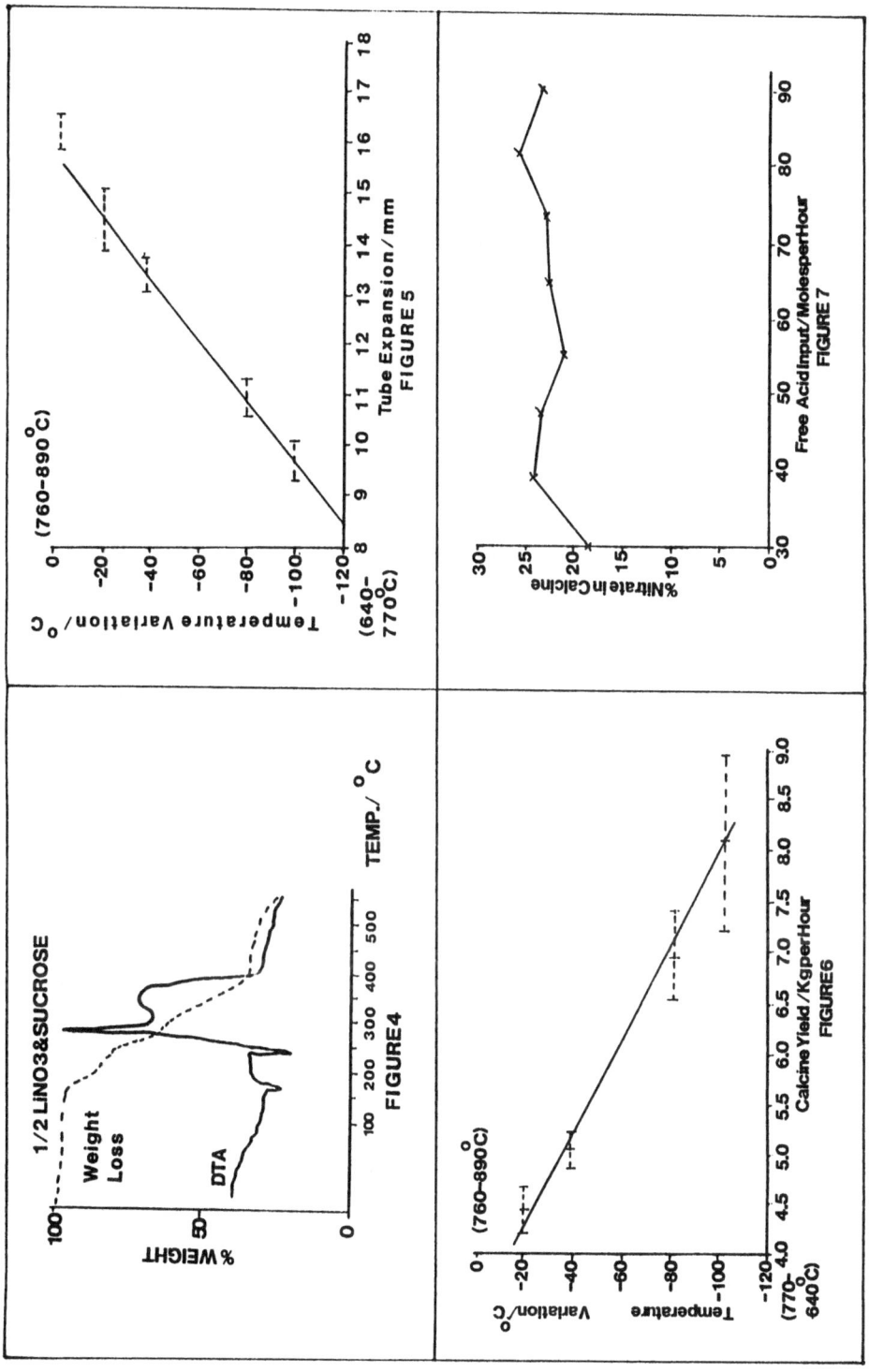

1/2 LiNO3 & SUCROSE

FIGURE 4

FIGURE 5

FIGURE 6

FIGURE 7

149

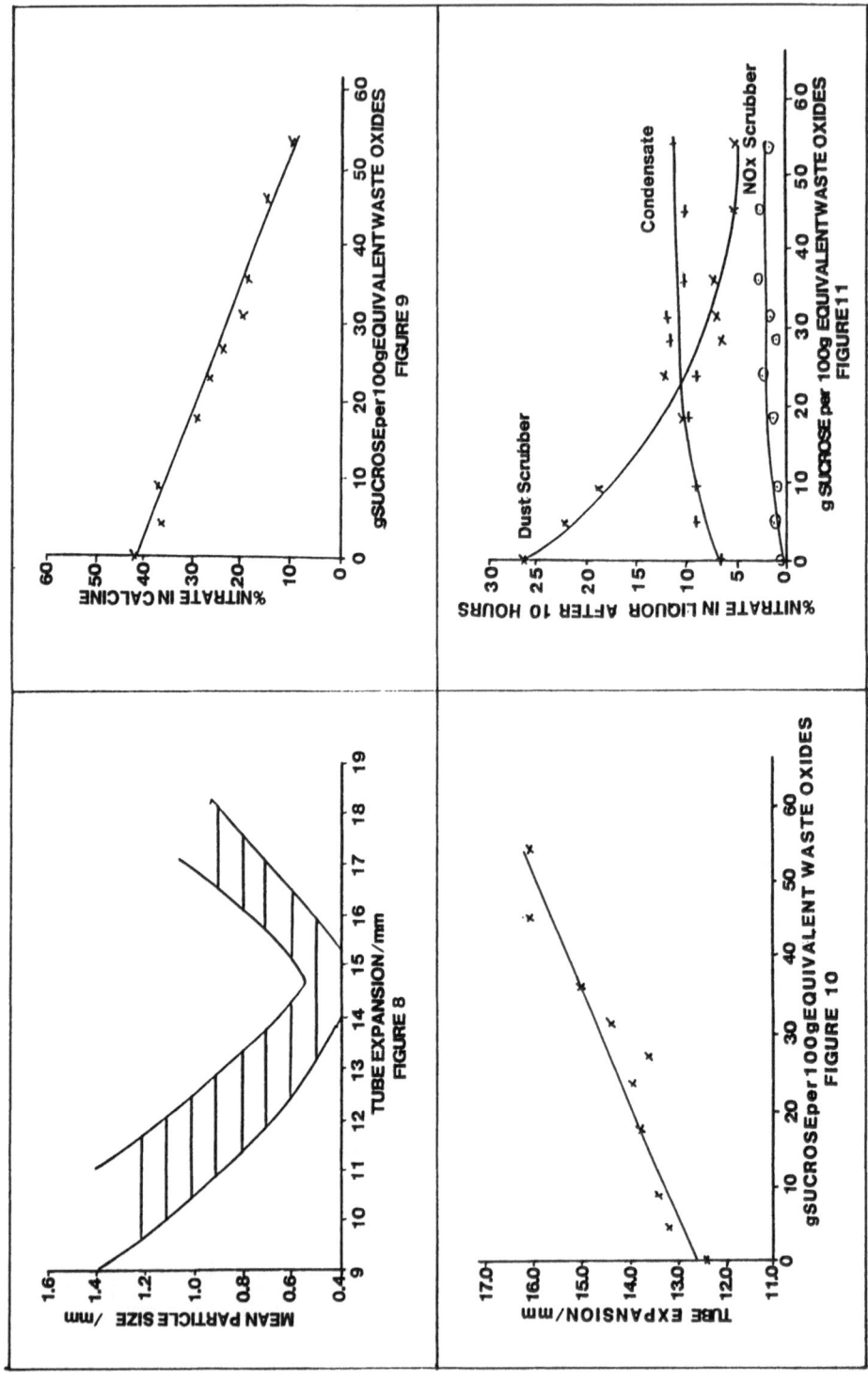

FIGURE 9

FIGURE 8

FIGURE 10

FIGURE 11

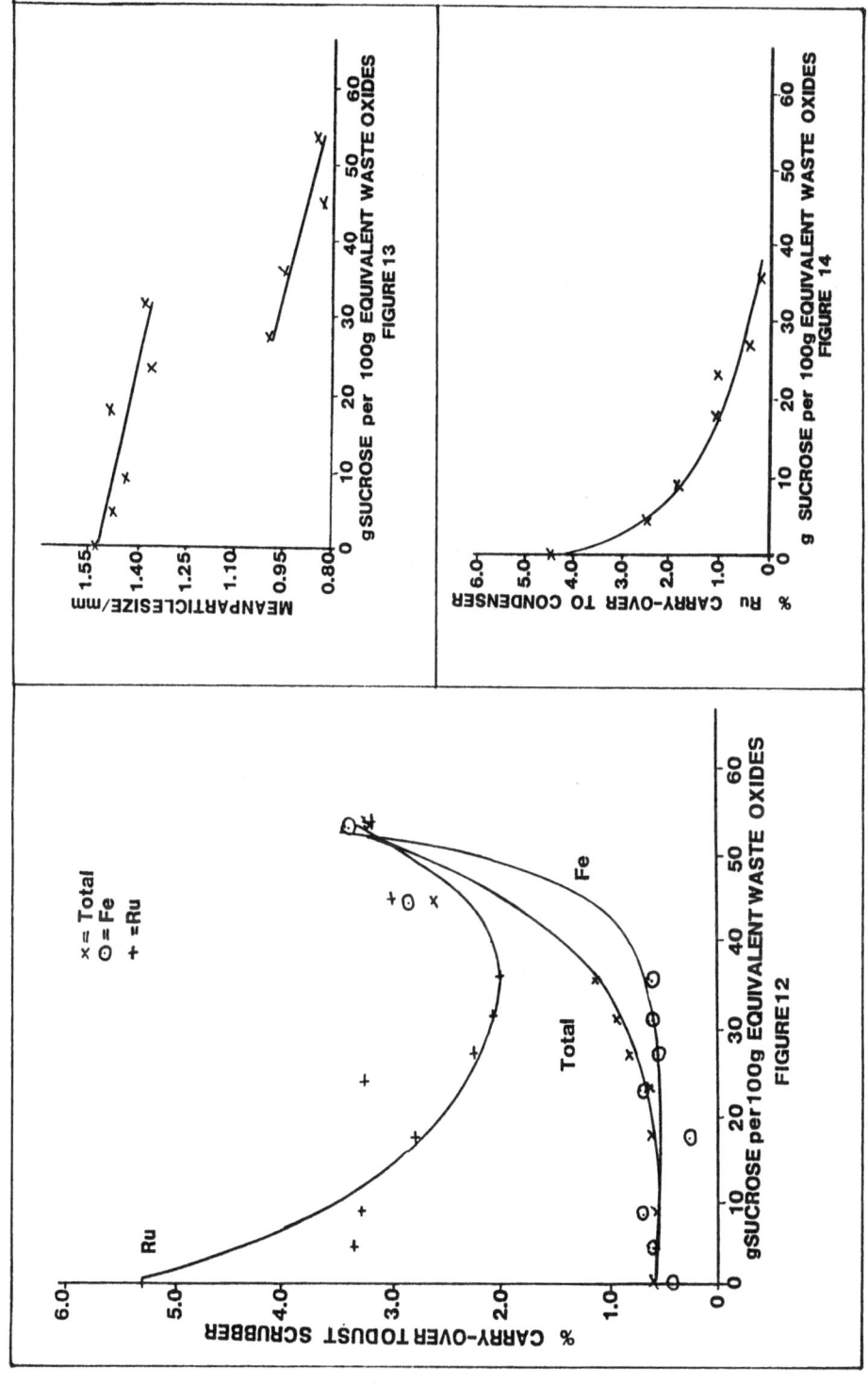

Figure 12
% CARRY-OVER TO DUST SCRUBBER
gSUCROSE per 100g EQUIVALENT WASTE OXIDES

x = Total
⊙ = Fe
+ = Ru

Figure 13
MEAN PARTICLE SIZE/mm
gSUCROSE per 100g EQUIVALENT WASTE OXIDES

Figure 14
% Ru CARRY-OVER TO CONDENSER
g SUCROSE per 100g EQUIVALENT WASTE OXIDES

REDUCTION OF Ru AND Tc VOLATILITY
DURING VITRIFICATION OF HLLW BY DENITRATION

A. JOUAN*, J.P. MONCOUYOUX* AND S. HALASZOVICH**

*Commissariat à l'Energie Atomique (CEA), Marcoule, France
**Kernforschungsanlage (KFA), Jülich, FRG

SUMMARY

Vitrification processes for high level wastes in nitric acid
solutions require drying, calcination and melting operations that
are conducive to the formation of oxidized compounds of certain
elements. These include ruthenium and technetium with the volatile
oxides RuO_4 and Tc_2O_7 corresponding to a high degree of oxidation.
In addition to selective traps, various methods are available for
limiting this volatilization to reduce the radioactive and techno-
logical hazards. One of these methods, predenitration, has been
investigated for many years, as a means of significantly reducing
the oxidation of nitrosyl ruthenium compounds produced above 130°C
during calcination. Another method involves the addition of reducing agents directly to the HLLW solution prior to calcination.
Extensive testing on lab-scale equipment and semi-industrial pilot
facilities has made it possible to define the optimum experimental
conditions capable of reducing the total ruthenium and technetium
volatility in industrial high level waste vitrification facilities
to 1.5% and 8%, respectively, of the feed solution activity.

1 INTRODUCTION AND GENERAL BACKGROUND

Among the elements produced by uranium fission, ruthenium, with its
complex chemistry, has been the subject of considerable work to determine
its behavior during reprocessing.

LWR fuel with a burnup of 33 000 $MWd \cdot t^{-1}$ contains 2220 grams of Ru
per metric ton of heavy metal. After cooling for 3 years, the Ru activity
accounts for 17% of the total fission product $\beta\gamma$ activity and for 21% of
the residual power.

Ruthenium is present in spent fuel solutions as nitrated complexes of
nitrosyl ruthenium in equilibrium. The general formula of these complexes
is the following |1|: $RuNO (NO_2)_x (NO_3)_y (OH)_{3-x-y} (H_2O)_{5-x-y}$.

Thermal decomposition of these complexes in an acid medium during
vitrification results in volatilization of ruthenium tetroxide (RuO_4)
which reduces to form a blue-black solid deposit of fine RuO_2 particles
inside the process lines in a self-catalyzing reaction. RuO_2 tends to
form preferentially between 150°C and 500°C |2|, and can plug system lines
or interfere with process operation if sufficient buildup occurs.

Technetium, an artificial β-emitter (0.29 MeV) forms at the rate of 836 grams per metric ton of heavy metal in LWR fuel irradiated to 33 000 $MWd \cdot t^{-1}$. All oxidation states from -1 to +7 are possible [3], but in nitric acid fuel solutions it is found as the pertechnetate anion $(TcO_4)^-$. Technetium constitutes a long-term radiation hazard because of its long half-life (2.13×10^5 years) and its high chemical mobility in geological systems.

In high-temperature processes such as vitrification, volatilization of technetium oxides is possible. The most stable oxide is the +7 valency Tc_2O_7. This oxide has a melting point of 119.5°C and a boiling point of 310.6°C. Four-valent TcO_2 has also been isolated: it sublimates above 900°C, and dismutes above 1100°C into Tc and Tc_2O_7.

2 METHODS FOR REDUCING VOLATILIZATION

Several methods are available for controlling volatilization during vitrification to limit the radioactivity hazard due to volatile elements and the potential fouling of process lines by solid ruthenium compounds.

2.1 Selective traps using the active properties of Fe_2O_3 to convert RuO_4 into RuO_2 [4].

2.2 Prior chemical denitration of the feed solution by addition of formaldehyde (HCHO) or formic acid (HCOOH).

Denitration with formaldehyde has been investigated by KFA at Jülich, and is illustrated by the following reactions [5]:

$$4 \ HNO_3 + HCHO \ \longrightarrow 2 \ NO_2 + CO_2 + 3 \ H_2O \tag{1}$$
$$4 \ HNO_3 + 3 \ HCHO \longrightarrow 4 \ NO + 3 \ CO_2 + 5 \ H_2O \tag{2}$$

Reaction (1) is applicable at high acidities (above 8 M HNO_3), while the second equation is valid at lower acidities.

The objective is to destroy all of the free nitric acid and fission product nitrates in the nitrosyl ruthenium complexes in order to limit subsequent NO_2 formation during calcination and vitrification, which would be conducive to the production and volatilization of RuO_4.

Denitration with formic acid has been tested by the CEA at Marcoule using the following reactions:

$$2 \ HNO_3 + 3 \ HCOOH \longrightarrow 2 \ NO + 3 \ CO_2 + 4 \ H_2O$$
$$2 \ HNO_3 + HCOOH \ \longrightarrow 2 \ NO_2 + CO_2 + 2 \ H_2O$$
$$2 \ HNO_3 + 4 \ HCOOH \longrightarrow N_2O + 4 \ CO_2 + 5 \ H_2O$$
$$2 \ HNO_3 + 5 \ HCOOH \longrightarrow N_2 + 5 \ CO_2 + 6 \ H_2O$$

2.3 Direct addition of reducing agents in the process stream to prevent oxidation of Ru to +8 valency and retain it as a nonvolatile compound.

Experiments of this type have been conducted at Marcoule in which a reaction was induced between the feed solution and sugar or formic acid directly in the calciner.

The denitration experiments at Jülich and Marcoule may be considered highly representative of reality since they were systematically conducted with actual reprocessing solutions containing nitrosyl ruthenium complexes.

In neither case was any unusual volatilization of ruthenium compounds noted during denitration. However, the denitration experiments at Jülich revealed a very slight amount of technetium in the distillate, but this was attributable to entrainment by aerosol formation rather than to volatilization |6|.

3 PROCEDURES FOR INVESTIGATING VOLATILITY IN VITRIFICATION

Research over the years on solidifying high-level liquid waste (HLLW) in Germany and in France has provided substantial additional data, and the results of various programs may be compared:

- laboratory results (KFA Jülich),
- industrial pilot batch process results: PIVER (France) and FIPS (FRG),
- industrial continuous process results: AVM and ATLAS (France).

3.1 Laboratory Tests

Laboratory tests were recently conducted at Jülich on 20 liters of simulated medium-level waste solutions mixed with actual radioactive solutions. After denitration, Ru, Cs and Sb were precipitated, separated and mixed as slurry with the glass frit. The resulting product was dried at 115°C and vitrified in an alumina crucible at 1170-1200°C.

3.2 PIVER Process

The PIVER pilot facility was used at Marcoule from 1969 to 1973 for 164 batch operations during each of which about 200 liters of HLLW were converted in a single reactor into 100 kg of glass through consecutive drying, calcination, vitrification and casting steps.

3.3 FIPS Process

The FIssion Product Solidification (FIPS) process involves prior denitration of the feed solution before adding glass frit slurry which is then drum-dried at 120-130°C before melting in an induction-heated metal crucible. The pilot facility is capable of processing 1 kg glass.h^{-1}

3.4 AVM and ATLAS

The French continuous vitrification process, based on consecutive calcination and vitrification steps in two separate units, is currently implemented in two different plants. The Marcoule Vitrification Facility (AVM) is an industrial setup processing actual fission product solutions at the rate of 36 l·h^{-1} to produce 15 kg of glass per hour. ATLAS is an experimental pilot facility of smaller throughput capacity (20 l·h^{-1} and 10 kg·h^{-1}) operated with simulated inactive solutions spiked with radioactive tracers.

The laboratory and FIPS experiments involve prior denitration whereas the PIVER and AVM facilities vitrify undenitrated feed solutions.

The ATLAS facility is capable of operating with or without prior denitration, and with the addition of reducing agents during calcination. The evaporator used for denitration in this unit has a 60-liter useful capacity, and has operated continuously with simultaneous process feeding and removal at the rate of 16 $1 \cdot h^{-1}$.

4 RUTHENIUM VOLATILIZATION

4.1 Ruthenium Volatilization in a Laboratory Experiment

A laboratory experiment was conducted on a denitrated solution with the composition indicated in Table I. Total ^{106}Ru activity was 8.7 mCi, of which 1.15 mCi was recovered in the offgas processing stream, representing a 13% volatilization.

TABLE I: Composition of Simulated MAW-S/12 Solution

Na	54.12	$g \cdot 1^{-1}$	Mn	0.08	$g \cdot 1^{-1}$
Al	0.23	$g \cdot 1^{-1}$	Mo	0.38	$g \cdot 1^{-1}$
Ca	1.5	$g \cdot 1^{-1}$	Na	0.18	$g \cdot 1^{-1}$
Cr	0.08	$g \cdot 1^{-1}$	Ni	0.08	$g \cdot 1^{-1}$
Cs	0.001	$g \cdot 1^{-1}$	Sr	0.001	$g \cdot 1^{-1}$
Cu	0.15	$g \cdot 1^{-1}$	Zn	0.15	$g \cdot 1^{-1}$
Fe	0.38	$g \cdot 1^{-1}$	Zr	0.08	$g \cdot 1^{-1}$
Mg	0.75	$g \cdot 1^{-1}$	HNO_3	1	$mole \cdot 1^{-1}$

4.2 Ruthenium Volatilization in the PIVER Batch Process

Over a 6-month operating period, the mean Ru volatilization recovered in the condensates was 15% |7-8|. Each batch involved approximately 800 curies of ^{106}Ru. The most significant Ru losses occurred during drying and at the beginning of the calcining operation up to 450°C. This substantiates the hypothesis that ruthenium volatilization begins near 130°C |3|.

4.3 Ruthenium Volatilization in the AVM Continuous Process

The first vitrification run in the AVM facility resulted in condenser recovery of 17% of the original Ru present |9|. The total volatilization loss from the calciner was 39% of the initial Ru quantity, of which 57% was retained by the dust scrubber.

4.4 Ruthenium Volatilization in the ATLAS Continuous Process

Table II indicates the test conditions in the ATLAS facility |10|.

TABLE II: ATLAS Test Parameters

TEST	REAGENT	QUANTITY OF SOLUTION PROCESSED	INACTIVE Ru CONCENTRATION $RuNO(NO_3)_3$	Ru ACTIVITY	TYPE OF SOLUTION
Without denitration or reducing additive	None	1500 l	$0.1 \ g \cdot l^{-1}$	$100 \ \mu Ci \cdot l^{-1}$	Table III
Prior denitration	Formic acid	5168 l	$0.1 \ g \cdot l^{-1}$	$200 \ \mu Ci \cdot l^{-1}$	Tables III & IV
Reducing additive in calciner	Sugar	6400 l	$0.1 \ g \cdot l^{-1}$	$200 \ \mu Ci \cdot l^{-1}$	Table III
	Formic acid	1190 l	$0.1 \ g \cdot l^{-1}$	$300-400 \ \mu Ci \cdot l^{-1}$	

TABLE III: Composition of a Simulated Waste Solution

K	$4.78 \ g \cdot l^{-1}$		Mn	$2.99 \ g \cdot l^{-1}$
Sr	$1.58 \ g \cdot l^{-1}$		Co	$0.34 \ g \cdot l^{-1}$
Ba	$2.54 \ g \cdot l^{-1}$		Ni	$2.07 \ g \cdot l^{-1}$
Zr	$6.55 \ g \cdot l^{-1}$		Fe	$10.38 \ g \cdot l^{-1}$
Ce	$4.73 \ g \cdot l^{-1}$		Na	$17.36 \ g \cdot l^{-1}$
La	$8.56 \ g \cdot l^{-1}$		Al	$5.45 \ g \cdot l^{-1}$
Pr	$2.06 \ g \cdot l^{-1}$		P	$0.52 \ g \cdot l^{-1}$
Nd	$7.16 \ g \cdot l^{-1}$			
Mo	$5.86 \ g \cdot l^{-1}$		HNO_3	$2 \ moles \cdot l^{-1}$

TABLE IV: Composition of a Simulated Waste Solution

K	$6.05 \ g \cdot l^{-1}$		Nd	$4.73 \ g \cdot l^{-1}$
Sr	$1.86 \ g \cdot l^{-1}$		Mo	$9.03 \ g \cdot l^{-1}$
Ba	$3.49 \ g \cdot l^{-1}$		Mn	$4.04 \ g \cdot l^{-1}$
Zr	$7.86 \ g \cdot l^{-1}$		Co	$0.69 \ g \cdot l^{-1}$
Ce	$10.26 \ g \cdot l^{-1}$		Ni	$2.44 \ g \cdot l^{-1}$
La	$5.61 \ g \cdot l^{-1}$			
Pr	$1.30 \ g \cdot l^{-1}$		HNO_3	2 moles·1-1

* Ru Volatilization without Denitration or Reduction

These operating conditions are representative of the AVM facility, in which the calcination additive is an organic reagent.

The quantity of Ru found in the condensate represents 20% of the initial Ru content. The total Ru volatilization loss from the calciner was 43%, of which 53% was retained by the dust scrubber.

* Ru Volatilization after Prior Denitration

Five different denitration test configurations (Table V) were investigated to determine the extent to which the degree of nitrate destruction by formic acid affects Ru volatilization. Radioactive ruthenium was present in the evaporator in each case.

TABLE V: Denitration Conditions

TEST No	TYPE OF SIMULATED SOLUTIONS	$\dfrac{\text{Moles HCOOH}}{\text{Moles NO}_3}$ IN EVAPORATOR	DENITRATED SOLUTION	
			$\text{moles} \cdot \text{l}^{-1}$ NO_3^-	$\text{moles} \cdot \text{l}^{-1}$ HNO_3
16	Table IV	1.53	N.D.	N.D.
19	Table IV	1.21	0.80	0
20	Table IV	1.98	0.65	0
22	Table IV	1.00	1.46	0.26
25	Table III	1.00	2.18	0

N.D. = not determined

The measured volatilization relative to the Ru quantities found in the dust scrubber, condenser and process lines is indicated in the table below.

TABLE VI: Volatilization Test Results with Prior Denitration

TEST No	Ru VOLATILIZATION OUTSIDE CALCINER (%)	Ru FOUND IN CONDENSER (%)
16	5.50	0.50
19	1.05	0.02
20	1.50	0.12
22	14.50	1.60
25	5.00	2.00

The amount of sugar used here (30 g) is effective for the solution indicated in Table III, but would probably have to be optimized for other types of solutions with higher acidity or nitrate content, since Ru volatilization increases as the sugar content drops.

Compared with prior denitration, the use of a sugar additive in the calciner is equally effective in reducing RuO_4 volatilization, and is easier to implement on an industrial scale.

5 TECHNETIUM VOLATILIZATION

5.1 Tc Volatilization at Lab Scale and in the FIPS Process

A series of tests conducted at Jülich made it possible to quantify Tc volatilization during calcination and vitrification |6|. A simulated feed solution and glass frit mixture was heated to 700°C and the volatilization was measured as a percentage of the initial Tc activity.

TABLE VIII: Temperature Effect on Tc Volatilization

TEMPERATURE (°C)	RESIDENCE TIME (min)	Tc VOLATILIZATION (%)	TOTAL WEIGHT LOSS (%)
250	30	Undetectable	1.7
300	30	Undetectable	2.6
350	30	Undetectable	3.5
400	40	Undetectable	4.5
450	60	Undetectable	5.3
525	60	0.1	5.8
600	60	1.1	6.2
675	60	10.0	6.5

Volatilization clearly begins at 525-550°C, and becomes significant at 600-700°C. From the weight loss percentages, it can also be seen that Tc volatilization is not a direct consequence of thermal denitration, since it becomes predominant after the nitrate destruction is largely completed.

In vitrification conditions, immediate and strong Tc volatilization was observed when either a simulated feed solution or the dried waste product were fed into a red-hot glass melt. The degree of volatilization seems to depend more on the residence time in contact with the molten glass than on the glass melt temperature. The test results are indicated in Table IX.

Volatilization to this extent results in condensation of Tc compounds forming shiny black solid deposits on the cooler portions of the process equipment.

It is clear from these results that Ru volatilization in the continuous process is considerably reduced by prior denitration of the feed
solution. In order to be effective, a maximum of the nitrates must be
destroyed. In test 22, for example, where the denitrated and vitrified
solution was slightly acid (0.26 M), substantial volatilization occurred.
In the non-acid solutions, the quantity of nitrates remaining in solution
affected the Ru volatilization percentage:

Test No 25: 2.18 moles NO_3/liter ---> 5% volatilization
Test No 20: 0.68 moles NO_3/liter ---> 1.5% volatilization

These results confirms that thermal nitrate decomposition during
calcination is favorable to Ru oxidation to the +8 valency and formation
of the volatile oxide RuO_4.

* Ru Volatilization with Formic Acid Added during Calcination

Formic acid is added to the undenitrated feed solution with an
$HCOOH/NO_3$ molar ratio of 0.28. Under these conditions, Ru volatilization
is 50-55% at the calciner outlet, and 8-12% in the condenser.

This procedure does not diminish ruthenium losses during calcination,
because the solution residence time in the calciner is too short to allow
the denitration reaction to occur. The formic acid subsequently distills
and reacts in the dust scrubber and condenser, where the nitric acidity
decreases.

* Ru Volatilization with Saccharose Additive

Sugar is usually added before calcination by mixing the feed solution
with a molar aqueous saccharose solution. The following table indicates
the proportions of saccharose added to the 200 $\mu Ci \cdot l^{-1}$ feed solution and
the Ru volatilization results obtained:

TABLE VII: Volatilization Test Results with Saccharose Additive

TEST No	SUGAR CONTENT OF FEED	TOTAL VOLATILIZATION	RUTHENIUM FOUND IN CONDENSATE
28	10 $g \cdot l^{-1}$	25.67%	11.64%
26	20 $g \cdot l^{-1}$	5.46%	1.06%
29	25 $g \cdot l^{-1}$	2.05%	0.33%
27	30 $g \cdot l^{-1}$	1.42%	0.19%

These findings show the remarkable effectiveness of sugar in reducing
Ru volatility: 30 g of saccharose per liter of simulated feed solution
were sufficient to diminish the total Ru loss outside the calciner from
43% (without reduction) to 1.5%. It is likely that the carbohydrates
released by hydrolysis of the saccharose are active enough to prevent the
oxidation of nitrosyl ruthenium complexes to RuO_4 during calcination.

TABLE IX: Technetium Volatilization during HLLW Vitrification

FEED RATE	FEEDING MODE	GLASS MELT TEMPERATURE (°C)	EXPERIMENTAL SETUP	VOLATILIZATION (%)
20 g over 1 h	Dry	1150	Lab-scale	43
20 g over 1 h	Dry	1150	Lab-scale	45
20 g over 1 h	Liquid	1150	Lab-scale	48
20 g over 2 h	Dry	1150	Lab-scale	52
20 g over 2 h	Liquid	1150	Lab-scale	58
2 kg over 1 h	Dry	1150	FIPS	61
2 kg over 2 h	Dry	1150	FIPS	59

5.2 Tc Volatilization in the ATLAS Facility

The only results available for technetium are those obtained during a continuous vitrification test with 30 grams of sugar added per liter of feed solution |10|; the solution composition is specified in Table III. The feed was calcined for 88 hours at the rate of 16 $l \cdot h^{-1}$, and a total of 1379 liters of solution were vitrified.

Technetium was introduced into the solution as sodium pertechnetate, with a ^{99}Tc activity of 200 uCi$\cdot l^{-1}$. The total Tc volatilization was 8%, and the activity recovered in the condenser represented 3% of the total initial ^{99}Tc activity.

These values are higher than for ruthenium, indicating that the mechanisms of the oxidation reactions leading to the formation of Tc_2O_7 are not the same as those resulting in oxidation of ruthenium to RuO_4. Moreover, the results at Jülich showed that Tc volatilization occurs at temperatures above 600°C, whereas RuO_4 losses are observed in the range from 130°C to 450°C.

6 CONCLUSION

After several years of work aimed at minimizing the entrainment of volatile ruthenium compounds in vitrification process facilities, the probable RuO_4 volatilization can be limited to 1.5% of the feed solution activity. This result may be obtained either by thorough predenitration of the feed solution or by calcination with a sugar additive. This low volatilization is desirable from the standpoint of the activity released in the process gas discharges before purification, and technologically in so far as it prevents fouling of process lines by solid powdery RuO_2 deposits.

Chemical reducing agents appear to be less effective with technetium, for which 3% volatilization is observed after dust separation. This factor must be taken into account in the vitrification offgas purification systems.

REFERENCES

|1| DIANA, J.J. Comportement du ruthénium lors du traitement des combustibles irradiés. Rapport CEA-R 4813 (1977).

|2| ORTINS DE BETTENCOURT, A., and JOUAN, A. Volatilité du ruthénium au cours des opérations de vitrification des produits de fission. Rapport CEA-R 3663 (1963).

|3| JONES, A.G., and DAVISON, A. The Chemistry of Technetium I, II, III and IV. Int J. Appl. Isot. vol 33, pp 867-874 (1972).

|4| BONNIAUD, R., JOUAN, A., LAUDE., F. and SOMBRET, C. Traitement des effluents gazeux dans les installations de vitrification de produits de fission. Seminar on Radioactive Effluents from Nuclear Fuel Reprocessing Plants. CEC-EUR 6076: Karlsruhe (Nov 22-25, 1977). Proc. pp 621-652.

|5| ODOJ, R., MERZ, E. and WOLTERS, R. Effect of Denitration on Ruthenium Volatilization. Scientific Basis for Nuclear Waste Management, vol 2, pp 911-917 (1980).

|6| LAMMERTZ, H., MERZ, E. and HALASZOVICH, S. Technetium Volatilization during HLLW Vitrification. Mat. Res. Soc. Symp. Proc. vol 44 (1985).

|7| BONNIAUD, R., LAUDE., F. and SOMBRET, C. Expérience acquise en France dans le traitement par vitrification des solutions concentrées de produits de fission. Colloque sur la gestion des déchets radioactifs résultant du traitement des combustibles irradiés. OCDE: Paris (Nov 27-Dec 1, 1972). pp 555-592.

|8| BONNIAUD, R. La vitrification en France des solutions de produits de fission. Nuclear Technology, vol 34, pp 449-460 (Aug 1977).

|9| COSTE, J.A., JOUAN, A.F., PAPAULT, C.C. and PORTEAU, C.M. Vitrification of High-Level Waste Solutions at Marcoule. European Nuclear Conference: Hamburg (May 1979).

|10| LAUDE, F. and CARTIER, R. Rapports d'exploitation de l'atelier ATLAS. Publications internes au CEA.

INDUSTRIAL APPLICATION OF DENITRATION OF HLLW
BY MEANS OF HCHO

C.BRESCHET - D.PAGERON (COGEMA, reprocessing division)
F.DRAIN - V.DECOBERT (SGN, process department)

Summary

Concentration of HLLW is an interim step between reprocessing and solidification process. The process using denitration by means of HCHO is composed of three main steps : receiving of solutions and feeding, concentration in a kettle type evaporator, and absorption of NOx. This process is operated semi-continuously : concentration is performed at constant level (with denitration and NOx absorption) and evaporator is emptied by batch to the storage. The industrial experience has been gained for nearly 30 years in three reprocessing plants. The safety of the chemical process reaches very high standard thanks to the appropriated operating mode and efficient process control. Only very few incidents have been recorded. No alternative to this process is envisaged today.

1. INTRODUCTION

1.1 Purpose of HLLW denitration

High level liquid waste for "PUREX" reprocessing plants are mainly composed of fission products dissolved in aqueous acid solution. This remaining solution, after separation of uranium and plutonium, contains more than 99% of the dissolved fission products, together with impurities from cladding materials, corrosion products, traces of unextracted plutonium and uranium, and most of the other transuranic elements.

These liquid wastes are concentrated, in order to minimize the storage volume. This also allows recovering of the nitric acid.

Because, for the long term, these liquid wastes should be converted to solid form, liquid storage can be considered as an interim step between reprocessing and solidification process.

In order to avoid problems of corrosion and precipitation in storage tanks, it is important to control the acidity of HLLW concentrated solutions. This goal can be attained by denitration.

1.2 Location of HLLW concentration

See figure 1.

FIGURE 1.
Location of HLLW concentration

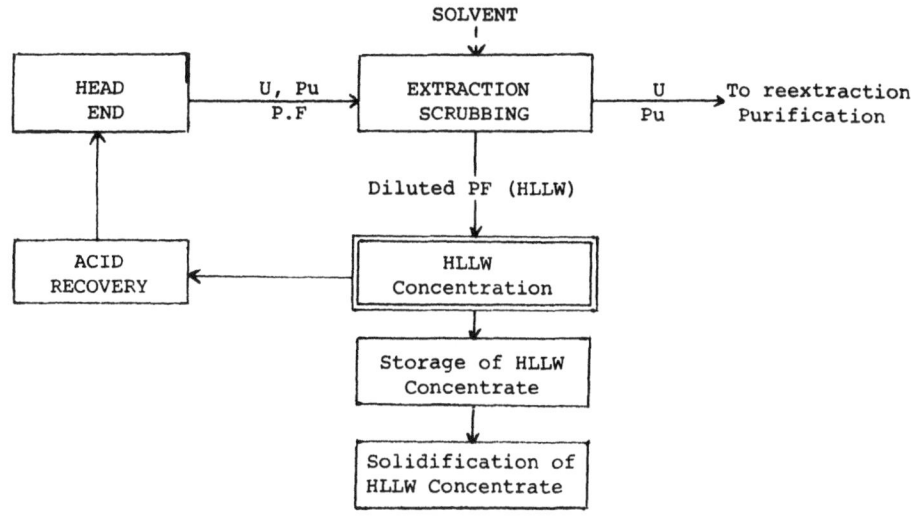

1.3 General principle of formic denitration using formaldehyde

. The global reaction of nitric acid reduction by formaldehyde gives the main following products :

$$CO_2, H_2O, NO, NO_2$$

The formaldehyde consumption depends mainly on the acidity in the evaporator.

In continuous denitration process, with a free acidity of about 2N in the evaporator, according to industrial experience, the mean consumption factor is 0.4 to 0.5 mole of formaldehyde per mole of destroyed acid in the solution.

. After denitration, the nitrous vapours are recombined.

The chemical reactions involved are :

in condenser : $NO_2 + \frac{1}{3} H_2O \longrightarrow \frac{2}{3} HNO_3 + \frac{1}{3} NO$

in absorption column : $\frac{4}{3} NO + \frac{2}{3} H_2O + O_2 \longrightarrow \frac{4}{3} HNO_3$

2 DESCRIPTION OF A TYPICAL HLLW CONCENTRATION UNIT

Diagram : see figure 2

A typical HLLW concentration unit, is composed of three main parts :
- Receiving of solutions - Feeding
- Evaporation
- NOx absorption

There may be several lines for evaporation or NOx absorption.

2.1 Receiving of solutions - Feeding

The solutions are received in one or several tanks.

The feed is sent to the evaporator :
- by means of a large flowrate transfering device, to fill the evaporator before denitration (steam jet for instance).
- by means of a controlled flowrate transfering device, to feed the evaporator during denitration.

2.2 Evaporation

The evaporator is kettle type, heated by external jacket (see table 1). Cooling is performed with chilled water, by means of the same jacket.

The evaporator is fitted with a decontamination column.

The gases at the outlet of the column go through a condenser, where recombination of nitrous vapors begins.

A part of recombined acid is refluxed to the evaporator.

2.3 NOx absorption

The NOx absorption is performed in a bubble-cap tray column.

There are two parts in this column :
- in the lower part, is performed the bulk of absorption.
- the upper part is a finishing zone.

. The non-condensed vapors from the condenser, and the oxygen necessary for the chemical reaction, are introduced at the bottom of the column.

A large flow of recombined acid from the bottom of the column is cooled and recycled to the top of the lower part, to remove the heat produced by the reaction. This heat can also be removed by a cooling jacket fitted on the column.

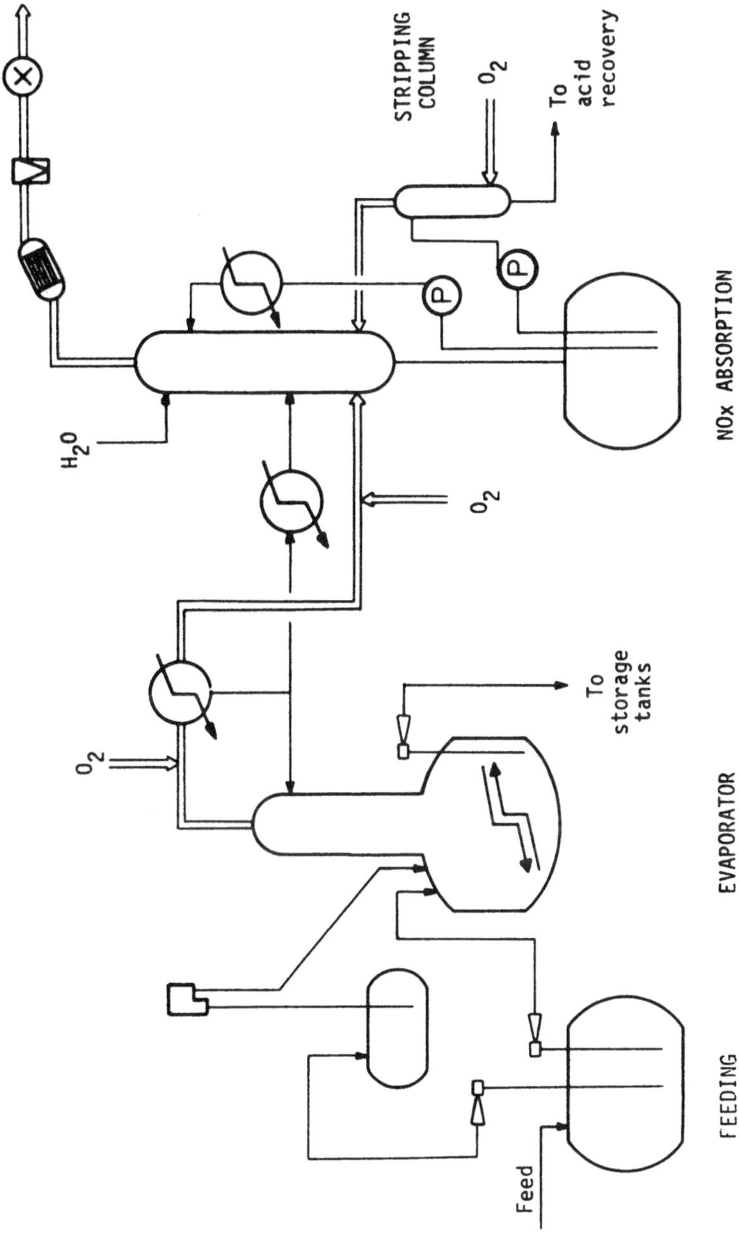

STRIPPING COLUMN

O_2

To acid recovery

NOx ABSORPTION

H_2O

O_2

O_2

To storage tanks

EVAPORATOR

FEEDING

Feed

FIGURE 2

The recombined acid from the condenser is also introduced in the lower part, after being cooled.
. In the finishing zone, remaining gases are contacted with fresh water feed at the top of the column.
The two phases of absorption can be performed in one bubble-cap tray column, or in two separate columns.
. The recombined acid is received in a tank. It is pumped and sent to the acid recovery, through a stripping column, the purpose of which is to remove dissolved nitrous vapours.

3 OPERATING PRINCIPLE

The unit is semi-continuously operated. A concentration cycle is composed of three main phases :
- Start-up : . filling of the evaporator
  . heating of the solution in the evaporator, without feeding up to boiling point
  . start-up of the reaction
- Concentration phase, with continuous feed at a controlled flowrate and continuous denitration
- End of cycle with . stop of feeding
  . total reflux
  . cooling of the solution in the evaporator
  . emptying of the evaporator
The nitrous vapors are recombined continuously.

3.1 Start-up
After reception of the solutions to be concentrated, the start-up of the unit is performed according to the following sequence :
- Filling of the evaporator with HLLW solution, up to the normal level of the evaporator, at a large flowrate,
- start-up (fresh water and recycling of acid) of the NOx absorption column,
- heating of the solution up to boiling temperature, with total refluxing,
- start-up of the oxygen feed to the NOx column,
- when the solution is boiling, and when the temperatures are stabilized, a small amount of sodium nitrite ($NaNO_2$) is sent in the evaporator to start the denitration reaction, and formaldehyde feeding is started,
- checking of the start-up of the denitration reaction by :
. analysis of distillates acidity,
. measurement of gaseous flowrate at outlet of the condenser (this flowrate increases because of nitrous vapours formation),
. follow-up of some parameters depending on the equipment (pressure, temperature,...),
- start-up of continuous feed of HLLW solution at its nominal value,
- adjustment of formaldehyde flowrate at its nominal value,
- control is set on automatic mode (if any).

3.2 Normal operation
Evaporator :
The liquid level in the kettle is held at a constant value by control either of the heating power, or of the feed flowrate.

All evaporation parameters are controlled :
- flowrate of formaldehyde,
- level, pressure and temperatures in the evaporator,
- acidity of distillates,
- acidity of concentrate is checked at least twice a day by sampling. The acidity is kept constant by control of formaldehyde flowrate.

NOx absorption : survey of oxygen and washing water flowrates, and of temperatures.

3.3 Shut-down
. Normal shut-down :
When the required concentration factor is reached :
- the HLLW solution feed and the formaldehyde feed are stopped,
- the solution is kept boiling with total refluxing during the time necessary to destroy remaining reagents (formaldehyde and by-products).
- then heating is stopped and the evaporator is cooled down progressively,
- a sampling is performed to check the characteristics of HLLW concentrate,
- the washing water and oxygen feeds to the NOx column are stopped,
- when the temperature in evaporator reaches about 60°C, the HLLW concentrate is transferred to the storage tanks by steam-jet.
. Unexpected shut-down :
Refer to part 5 - Safety of process.
REMARK : The concentration factor is generally very high (about 20) and it is easily followed by the ratio between volume of feeding solution and volume in the evaporator. Concentration phase lasts several days.

4 INDUSTRIAL EXPERIENCE

The denitration of HLLW by formaldehyde has been performed on an industrial scale for nearly 30 years. SGN has designed several concentration units for Marcoule and La Hague plants in France and for Tokai-Mura plant in Japan.

These units are characterized by :
. Acidity of HLLW concentrate of 1 to 3N, which allows the highest concentration factor while minimizing solids precipitation.
. High purification factor : thanks to the denitration process it is possible to implement efficient decontamination column with high washing flowrate. The purification factor is generally comprised between 10^6 and 10^8, depending of the type of decontamination column.
. High concentration factor which is not limited by the process itself. It depends only of the fuel type and cooling time and is limited either by specific heat of the solution or by the salt content.
. Usefull volume of evaporators ranges from 3 to 10 m³
Main characteristics of industrial units : refer to table 1

NAME	START-UP	NUMBER OF EVAPORATORS	EVAPORATION RATE	HEATING FLUID	REMARKS
EXISTING UNITS					
MARCOULE	02-1959 (shut-down in 05.1965)	1 + 1 spare	850 kg/h	Thermofluid	
	05.1965	1 + 1 spare	600 kg/h	Thermofluid	
LA HAGUE	1965 (june 1966 with HLLW)	3	350 kg/h	Superheated water	1 absorption column
	1985	3	1100 kg/h	Superheated water	1 absorption column
TOKAI-MURA	1974	1 + 1 spare	450 kg/h	steam	1 absorption column
FUTURE UNITS					
UP3	1988	3	1000 kg/h	Superheated water	3 absorption column (3 independent lines)
UP2 800	1991	3	1000 kg/h	Superheated water	"

5 SAFETY OF THE PROCESS

The nuclear safety linked to the very high activity level in this unit is not discussed here. This part only deals with safety linked to use of formaldehyde in an industrial unit.

Chemical hazard :

When formaldehyde is introduced in the solution to be denitrated, an induction period precedes the reaction. The starting of the reaction has to be well surveyed to avoid an accumulation of formaldehyde and over pressure in the evaporator caused by sudden and strong start-up of the reaction.

The induction period depends on the temperature, the acidity of the solution, and the concentration of nitrous ions in the solution.

The duration of the induction period is minimum at boiling point. It decreases when the acidity increases.

Nitrous ions are introduced under the form of $NaNO_2$. The concentration of nitrous products necessary to decrease significantly the induction period is 100 to 200 mg/l in the evaporator. The starting of denitration is ensured by this addition of $NaNO_2$.

Operating conditions :

In the presence of high level liquid waste, the reaction must be under control at any time; this goal is attained thanks to the following conditions :

- A temperature higher than 95°C in the evaporator when the reagents are sent (if temperature decreases below 95°C, the formaldehyde feed is stopped).

- Formaldehyde is introduced at a continuous flowrate; change of its feed rate is always performed by small steps only, in order to avoid accumulations of formaldehyde.

- The gaseous reaction products from the evaporator are quickly exhausted.

- The free acidity of the solution is kept around 2N.

Unexpected shut-down

These above conditions have to be respected both for start-up of the reaction, and during concentration. A stop of the reaction could happen with :

. a decreasing of temperature,
. a stop of formaldehyde feed,
. a too low acidity of the solution.

When an unexpected shut down happens (for instance shut down of electricity, with stopping of formaldehyde flowrate) :

- if the shut down last less than 5 mn, the reaction is started up again, without sodium nitrite addition,

- if the shut down lasts more than 5 mn, total refluxing is established, to destroy reagents before to restart the reaction using same procedure as for normal start (refer to part 3).

The safety of industrial units is very good, thanks to :

- operating mode qualified by the very long time experience,
- instrumentation :

. Control of the temperature, and start-up forbidden or automatic shut down when temperature decreases (stop of formaldehyde flowrate),

. Two different measurement devices for formaldehyde flowrate (with totalisation of flow),

. Control of the gaseous flowrate outside of evaporator,

. Control of all usual parameters in evaporator and decontamination column (pressure, level, density,...).

- Analytical measurments :
. acidity of the concentrated solution checked at least twice a day,
. Acidity of distillates checked during the induction period and during the concentration step (twice a day).
- Control of the reaction start-up is perfectly ensured by the above mentionned conditions and controls.

Incidents
- Marcoule, 1964
At the end of a concentration cycle, a problem happens on the reflux line (lost of flowrate). The acidity in the concentrate was about 5N. Heating was stopped and evaporator cooled down immediately. During the cooling phase an overpressure occurred : it was caused by violent reaction of remaining reducing agents in the solution. The consequences were limited to irradiation in some inactive area through lifting of active solutions in dip tubes.

Investigations and tests have been conducted following this incident. Two measures have been taken to avoid renewal of such incident :

- Acidity in the concentrate is now limited to 3N at maximum,

- After formaldehyde feed stop it is always compulsory to keep the solution boiling for some time, before cooling, in order to destroy completely the reagents and by products present in concentrate solution.

- Marcoule, 1975
An overpressure occured during the start-up of denitration reaction leading to similar consequences as in 1964. The solution to be concentrated was exceptionnally containing some hydrazine and it was found that the quantity of $NaNO_2$ introduced was not sufficient to destroy all hydrazine. Therefore there was not enough nitrous products to ensure a quick start-up of denitration reaction. However it must be noticed that the processing of such type of solution (which corresponds to very specific operation) is quite occasionnal. Such conditions are never met during LWR processing.

- LA HAGUE 1981
Following a general failure of electricity (including emergency power) all instrumentation was completely lost for several hours. The heating and reagents feed were stopped. Temperature decreases to about 60°C. Before to restart the reaction evaporator was kept boiling with total reflux for 5 hours. No incident occured, which shows the great safety of the process.

6 CONCLUSION

Almost 30 years of industrial experience have clearly demonstrated the advantages of this process.

- Very low corrosion risks thanks to the low acidity in the evaporator, and reducing conditions. Up to now no failure due corrosion has ever been recorded.

- High concentration factor achievable particularly thanks to the acidity control.

- Good purification factor thanks to the possibility to implement efficient gas washing.

- Although the chemical reactions involved present some potential hazards, the safety of the process is very high thanks to the well established operating mode and to the adequate process control.

The few recorded incidents have never lead to significant consequences either for environment or for operating people.

The performances and safety of this process being considered as very good, no alternatives are envisaged today.

CONCLUSIONS AND OPINION OF THE SCIENTIFIC SECRETARIES

1. GENERAL

Basically denitration of radioactive liquid waste is performed batchwise by feeding formic acid or formaldehyde into a nitric waste solution at boiling temperature or vice-versa. Although sugar and ethanol have been proposed as alternatives, their respective advantages over formic acid and formaldehyde (lower cost for sugar and better efficiency for ethanol) do not appear, in most cases high enough to counteract their inherent drawbacks (longer reaction time for sugar and high flammability in case of ethanol). The only application for which sugar is definitely superior deals with denitration of high level waste during vitrification (in the rotary kiln according to the AVM process) which enables a significant reduction of Ru and Tc volatilities. However, for all the other applications, so far, formic acid and formaldehyde remain the most suitable reagents for denitrating radioactive liquid waste.

Even if the knowledge of the fundamental mechanism of denitration reactions is not really necessary to conduct a denitration process - as demonstrated by the industrial application of denitration for many years - this could be helpful for optimizing the operating conditions particularly in case of exhaustive denitration of radioactive liquid waste where interfering reactions with some chemical components are bound to occur. In such a context, mathematical modelling of the evolution of denitration in time would be worthwhile taking into account the different basic parameters acting on the induction time and the denitration rate (initial nitric acidity, concentration of the reacting components, salt concentration, feeding rate, temperature, pressure variations, denitrator design ...).

2. PARTIAL DENITRATION OF RADIOACTIVE LIQUID WASTE

This is mainly applied to high level liquid waste for three purposes: concentration before storage and pretreatment prior to vitrification and possibly to actinide separation.

2.1. Concentration of high level liquid waste

In order to prevent corrosion of storage tanks, high level liquid waste is usually denitrated contemporaneously with its concentration keeping the acidity within 1-3N HNO_3 range. Since this application has reached the industrial scale for a few decades, the safety and the reliability of the process are no more questionable. Because the waste must be kept strongly acidic, the denitration/concentration process consists of feeding formaldehyde into boiling high level liquid waste.
It must be pointed out that under the operating conditions chosen (nitric acidity higher than 1N HNO_3, salt content increasing and high feeding rates) fast denitration rates are favoured. Therefore, unless the radioactive waste contains nitrous acid scavengers, all the best conditions are met for denitration to proceed smoothly.

Although denitration with formaldehyde has not been subject to much basic research, more thorough investigations in this area do not appear very worthwhile.

2.2. Pretreatment of high level liquid waste prior to vitrification

According to the basic concept adopted in F.R.G. for vitrification of high level liquid waste, the implementation of a denitration step upstream the ceramic melter could be advantageous for reducing the off-gas corrosiveness, the ruthenium and technetium volatility as well as the calcining temperature of the fission products. Since only acidity lowering from 5N to 1.5N HNO_3 is aimed at, denitration by feeding formaldehyde into the waste is envisaged. Therefore the experience gained on concentration of high level liquid waste (see previous section) is straightforward transposable without any need for further research.

2.3. Pretreatment of high level liquid waste before actinide partitioning

Denitration of high level liquid waste for actinide partitioning purposes differs according to the type of process so far envisaged. For the process based on solvent extraction (TBP2 process), denitration is only intended to lower the nitric acidity of concentrated high level liquid waste down to about 1N HNO_3. Even if the flow-sheet proposed considers denitration with formic acid, formaldehyde could be used as well, so that no further development in this area is needed. The only aspect which could deserve more investigations concerns the influence of the operating conditions chosen for denitration on the actinides sorption on the denitration/concentration precipitate.

With respect to the process dealing with co-precipitation of actinides along with rare earth oxalates, denitration of high level liquid waste aims at reaching a pre-determined acidity around 0.2M HNO_3. Although again the denitration step deals with addition of the waste solution into boiling concentrated formic acid, it does not seem there is a major counter-indication to use formaldehyde. The fact that a pre-determined nitric acidity is aimed at within a rather narrow concentration range could require a more stringent control of the operating conditions fixed for denitration and a follow up of the reaction. Regarding this, the adoption of the opposite procedure for denitrating (formic acid or formaldehyde feeding into boiling high level waste) could be advantageous especially as a device for an in-situ control of the evolution of the acidity in the reacting mixture by means of conductivity measurements has been set-up by the JRC-Ispra. In this case, no accurate determination of the initial acidity of the waste solution would be needed. However such a change is advisable as far as no red-oil formation is expected to take place in the waste (see section 5).

Under the operating conditions envisaged so far (waste feeding into boiling formic acid), oscillations of the reacting mixture attributed to oxalic acid are sometimes recorded. Moreover, in order to protect oxalic acid destruction via nitrous acid, it seems that some iron should be present in the waste solution. Therefore, it appears worthwhile to perform larger denitration experiments so that the extent of the problems encountered can really be appraised. On the other hand, since this denitration must be carefully controlled, some more basic investigations concerning the mechanism of the denitration reaction itself might be useful.

173

3.   EXHAUSTIVE DENITRATION OF RADIOACTIVE LIQUID WASTE

Complete destruction of nitric acid requires an excess of formal-dehyde or formic acid with respect to the stoichiometry of the various basic reactions involved. Since both organic reducing agents react simi-larly with nitric acid, the choice of the procedure to be applied (feeding of waste solution into the excess of reducing agent or vice-versa) - is dictated by the requirements in matter of off-gas composition. Exhaustive denitration of radioactive liquid waste has not yet reached the industrial scale for the different applications envisaged so far: pretreatment of high level liquid wastes for actinide partitioning and volume reduction of medium level liquid waste (reprocessing concentrate). While this situation results from current priorities in radioactive waste management, it must be underlined that on a technical standpoint, all the problems linked to exhaustive denitration have not yet been completely solved or clarified as it is explained further.

3.1.   Pretreatment of high level liquid waste for actinide partitioning

In a proposed flow-sheet for actinide partitioning based on solvent extraction (the HDEHP 1 process) exhaustive denitration of high level liquid waste is performed by metering the waste solution into an excess of formic acid. As a result, the pH achieved lies within the 1.5 - 2.5 range depending on the excess of formic acid. As far as denitration is con-cerned, this process requests no stringent control of the operating conditions since variations of the final concentration of formic acid does not much affect the pH. The critical point is to make sure that enough noble metals are present in the high level liquid waste to promote actinide solubilisation. If the lab-scale experiments highlighted some detrimental effects of noble metals (occurrence of oscillations in the reacting mixture) and even slight under pressure (slow-down of the reac-tion) on the denitration performance, it is now believed that both pheno-mena are due to the low feeding rates adopted. In all likelihood, if the process had been tested on a much larger scale with higher feeding rates, the impact of both parameters (noble metals concentration and pressure variations) would have been much less pronounced. However, experimental confirmation of this statement is desirable prior to draw any definite conclusion.

On the basis of some experimental results, the exhaustive denitra-tion of high level liquid waste by means of formaldehyde instead of formic acid seems to behave similarly except for the off-gas composition which is dominated by NO/NO_2 because of the procedure used (formaldehyde feeding into boiling waste).

3.2.   Volume reduction of medium level liquid waste (reprocessing concentrate

Reprocessing concentrates are currently managed by direct cementa-tion prior to transportation and disposal. As these must be neutralised before cementation, replacement of sodium hydroxide addition by denitra-tion is expected to reduce the volume of the waste products up to 30%. However such an operation is potentially economically attractive only if no noticeable amounts of secondary liquid waste are generated as those arising during the off-gas treatment. This seems achievable when the waste solution is fed into an excess of boiling formic acid (for reasons indicated further, feeding into an excess of commercial formaldehyde is not recommended) since, in this case, nitric acid is expected to be primarily reduced into nitrogen and mainly nitrogen protoxide which can be released into the atmosphere. However, regarding the off-gas composition resulting from similar denitration tests, the results achieved at KFA-Jülich and KfK-Karlsruhe proved to be quite divergent. A comparison between the respective operating conditions revealed that the recorded discrepancies have to be attributed to the difference of feeding rates; low feeding rates of waste solution favouring the reduction of HNO_3 into N_2/N_2O whereas higher feeding rates give rise to more NO/NO_2. This results from the redox potential of the reacting mixture which increases with HNO_3 feeding rate because denitration does not proceed instantaneously. Accordingly reduction of nitric acid is mainly limited to NO/NO_2 as shown by the basic reactions. On the other way round, if the feeding rate is levelled with the denitration rate, the reacting mixture is kept reducing and N_2O/N_2 are generated. Unfortunately, the adoption of low feeding rates may make the denitration rate more sensitive to small variations of the operating conditions and the waste composition. Therefore, it is thought that more investigations at the pilot or industrial scale would be helpful to appraise the real volume reduction factor which can be expected from denitration of reprocessing concentrates.

If a quantitative reduction of HNO_3 into N_2/N_2O proved to be ineffective on an industrial scale, it may be thus advisable to take advantage of the device developed by ENEA-Casaccia (catalytic reduction of NO/NO_2 into N_2) to complete the reduction.

4. REDUCTION OF RUTHENIUM AND TECHNETIUM VOLATILITY DURING THE VITRIFICATION OF HIGH LEVEL WASTE

Whatever the organic reducing agent used (formaldehyde, formic acid or sugar), denitration of high level liquid waste during vitrification entails an important reduction of ruthenium and technetium volatility. However, in case of the AVM vitrification process, addition of sugar to high level liquid waste just prior the rotary kiln enables to attain two significant improvements at the same time. First, the ruthenium and technetium volatility can be reduced down to 1.5 and 8% respectively and second the calcinate produced is very crumbly preventing thus the risk of lumps formation in the calciner. The procedure is now well defined and does not need any further research.

5. SAFETY OF THE DENITRATION PROCESS

Provided that a thorough chemical analysis of the waste solutions to be denitrated is performed (in order to make sure that no nitrous acid scavengers are present) and that temperature and feeding rates are under control, the denitration process presents a high degree of safety. Because commercial formaldehyde contains some methanol, it is recommended to feed it into boiling waste solution rather the other way round, otherwise the induction time is longer and flammable gases are generated when the reaction starts. However if some TBP degradation products are present in the waste solution this procedure might be no longer advisable due to possible "red-oil" formation. In this latter case, contemporaneous feeding of formaldehyde with waste solution into the denitrator - as practised at the industrial scale for concentrating high level waste - appears to be the safest way to proceed. This latter statement also applies to formic acid which over formaldehyde presents in addition the advantage of being used as a pure product and thereby preventing the occurrence of troublesome reactions with impurities when the waste solution is fed into it.

However it must be stressed that transportation or storage of denitrated liquid waste has to be carried out cautiously taking care that the respective vessels are always ventilated due to the fact that mixtures of nitric acid and formaldehyde/formic acid may suddenly react, even at room temperature when they are stored in sealed tanks or containers, because of the positive effect of pressure on the denitration rate.

At last, since some amounts of formic acid/formaldehyde are often carried over in the acidic condensate resulting from denitration the management of this secondary liquid stream needs to be carefully controlled in order to avoid any further untimely denitration.

L. Cécille   S. Halaszovich

BOGUSLAWSKI Z.

F.J. GATTYS
Verfahrenstechnik GmbH
Frankfurter Str. 168-176
D - 6078 NEU ISENBURG

BRESCHET Ch.

COGEMA-Marcoule
B.P. N° 170
F - 30200 BAGNOLS SUR CEZE

CECILLE L.

C.E.C.
200, rue de la Loi
B - 1049 BRUSSELS

COURTOIS Ch.

C.E.A.
C.E.N. Cadarache
DRDD/SEDFMA
B.P. N° 1
F - 13108 ST PAUL LEZ DURANCE CEDEX

DECOBERT V.

S.G.N.
1, rue des Hérons
F - 78184 ST QUENTIN YVELINES CEDEX

DIX S.

K.F.A. - Jülich
I C T
Postfach 1913
D - 5170 JULICH

DONATO A.

E.N.E.A.
CRE Casaccia
S.P. Anguillarese, 301
I - 00100 ROMA

DRAIN F.

S.G.N.
1, rue des Hérons
F - 78184 ST QUENTIN YVELINES CEDEX

DWORSCHAK H.

C.E.C.
Joint Research Centre
Ispra Establishment
I - 21020 ISPRA (Va)

EID C.

C.E.C.
200 rue de la Loi
B - 1049 BRUSSELS

FATHO K.	K.A.H. Im Breitspiel 7 D – 6900 HEIDELBERG 1
FREUND P.	D.W.K. Hamburger Allee 4 D – 3000 HANNOVER 1
FUCHS R.	F.J. GATTYS Verfahrenstechnik GmbH Frankfurter Str. 168–176 D – 6078 NEU ISENBURG
GEENS L.	S.C.K./C.E.N. Boeretang 200 B – 2400 MOL
GOMPPER K.	K.f.K. Karlsruhe I N E Postfach 3640 D – 7500 KARLSRUHE 1
GUERIN V.	C.E.A. C.E.N. Cadarache SEDMA SETED B.P. N° 1 F – 13108 ST PAUL–LEZ–DURANCE CEDEX
HAHN H.	D.W.K. Hamburger Allee 2–4 D – 3000 HANNOVER 1
HALASZOVICH S.	K.F.A. – Jülich I C T Postfach 1913 D – 5170 JULICH
HARMS R.	K.F.A. – Jülich ICH–5 Postfach 1913 D – 5170 JULICH
HERBRECHTER D.	K.A.H. Im Breitspiel 7 Postfach 103420 D – 6900 HEIDELBERG 1
HUMBLET L.	Belgoprocess Gravenstraat B – 2480 DESSEL
HUNT B.	C.E.C. Joint Research Centre Ispra Establishment I – 21020 ISPRA (Va)

JERVIS M.C.	B.N.F.L. Sellafield Works SEASCALE Cumbria UK – CA20 1PG
KELM M.	K.f.K. Karlsruhe I N E Postfach 3640 D – 7500 KARLSRUHE 1
KLONK H.	D.W.K. Hamburger Allee 2-4 D – 3000 HANNOVER 1
KRÄMER H.	K.f.K. Karlsruhe P W A Postfach 3640 D – 7500 KARLSRUHE 1
KROEBEL H.	K.f.K. Karlsruhe Postfach 3640 D – 7500 KARLSRUHE 1
LECOMTE M.	C.E.A. C.E.N. Fontenay-aux-Roses DGR/SEP/STU B.P. N° 6 F – 92265 FONTENAY-AUX-ROSES
MERZ E.	K.F.A. – Jülich I C T Postfach 1913 D – 5170 JÜLICH
MONCOUYOUX J.P.	C.E.A. C.E.N. – Valrho B.P. N° 171 F – 30205 BAGNOLS SUR CEZE CEDEX
ODOJ R.	K.F.A. – Jülich I C T Postfach 1913 D – 5170 JÜLICH
OSER B.	K.f.K. Karlsruhe I N E Postfach 3640 D – 7500 KARLSRUHE 1
PIETRELLI L.	E.N.E.A. CRE Casaccia SP Anguillarese, 301 I – 00100 ROMA

RICCI G.   E.N.E.A.
                                      CRE Casaccia
                                      S.P. Anguillarese, 301
                                      I - 00100 ROMA

SCHULENBERG T.   Dornier System GmbH
                                      Postfach 1360
                                      D - 7990 FRIEDRICHSHAFEN

SIMON R.   C.E.C.
                                      200, rue de la Loi
                                      B - 1049 BRUSSELS

STENERSEN F.   K.A.H.
                                      Im Breitspiel 7
                                      Postfach 103420
                                      D - 6900 HEIDELBERG 1

VASSALLO G.   C.E.C.
                                      Joint Research Centre
                                      Ispra Establishment
                                      I - 21020 ISPRA (Va)

VIDA J.   K.f.K. Karlsruhe
                                      I H C
                                      Postfach 3640
                                      D - 7500 KARLSRUHE 1